GREEN SCHOOLS

ATTRIBUTES FOR HEALTH AND LEARNING

Committee to Review and Assess the
Health and Productivity Benefits of Green Schools

Board on Infrastructure and the Constructed Environment

Division on Engineering and Physical Sciences

NATIONAL RESEARCH COUNCIL
OF THE NATIONAL ACADEMIES

THE NATIONAL ACADEMIES PRESS
Washington, D.C.
www.nap.edu

THE NATIONAL ACADEMIES PRESS 500 Fifth Street, N.W. Washington, DC 20001

NOTICE: The project that is the subject of this report was approved by the Governing Board of the National Research Council, whose members are drawn from the councils of the National Academy of Sciences, the National Academy of Engineering, and the Institute of Medicine. The members of the committee responsible for the report were chosen for their special competences and with regard for appropriate balance.

This study was supported by a Master Services Agreement between the National Academy of Sciences and the Massachusetts Technology Collaborative (awarded November 2004); Grant 1906 between the Barr Foundation and the National Academy of Sciences (awarded September 2004); and funding from the Connecticut Clean Energy Fund (awarded April 2005), the Kendall Foundation (awarded March 2005), and the U.S. Green Building Council (awarded February 2005). Any opinions, findings, conclusions, or recommendations expressed in this publication are those of the author(s) and do not necessarily reflect the views of the organizations or agencies that provided support for the project.

International Standard Book Number-13: 978-0-309-10286-5
International Standard Book Number-10: 0-309-10286-3

Copies of this report are available from the Board on Infrastructure and the Constructed Environment, National Research Council, 500 Fifth Street, N.W., Room 967, Washington, DC 20001; (202) 334-3376.

Additional copies of this report are available from the National Academies Press, 500 Fifth Street, N.W., Lockbox 285, Washington, DC 20055; (800) 624-6242 or (202) 334-3313 (in the Washington metropolitan area); Internet, http://www.nap.edu.

Printed and bound in Great Britain by Marston Book Services Limited, Oxford

THE NATIONAL ACADEMIES

Advisers to the Nation on Science, Engineering, and Medicine

The **National Academy of Sciences** is a private, nonprofit, self-perpetuating society of distinguished scholars engaged in scientific and engineering research, dedicated to the furtherance of science and technology and to their use for the general welfare. Upon the authority of the charter granted to it by the Congress in 1863, the Academy has a mandate that requires it to advise the federal government on scientific and technical matters. Dr. Ralph J. Cicerone is president of the National Academy of Sciences.

The **National Academy of Engineering** was established in 1964, under the charter of the National Academy of Sciences, as a parallel organization of outstanding engineers. It is autonomous in its administration and in the selection of its members, sharing with the National Academy of Sciences the responsibility for advising the federal government. The National Academy of Engineering also sponsors engineering programs aimed at meeting national needs, encourages education and research, and recognizes the superior achievements of engineers. Dr. Wm. A. Wulf is president of the National Academy of Engineering.

The **Institute of Medicine** was established in 1970 by the National Academy of Sciences to secure the services of eminent members of appropriate professions in the examination of policy matters pertaining to the health of the public. The Institute acts under the responsibility given to the National Academy of Sciences by its congressional charter to be an adviser to the federal government and, upon its own initiative, to identify issues of medical care, research, and education. Dr. Harvey V. Fineberg is president of the Institute of Medicine.

The **National Research Council** was organized by the National Academy of Sciences in 1916 to associate the broad community of science and technology with the Academy's purposes of furthering knowledge and advising the federal government. Functioning in accordance with general policies determined by the Academy, the Council has become the principal operating agency of both the National Academy of Sciences and the National Academy of Engineering in providing services to the government, the public, and the scientific and engineering communities. The Council is administered jointly by both Academies and the Institute of Medicine. Dr. Ralph J. Cicerone and Dr. Wm. A. Wulf are chair and vice chair, respectively, of the National Research Council.

www.national-academies.org

COMMITTEE TO REVIEW AND ASSESS THE HEALTH AND PRODUCTIVITY BENEFITS OF GREEN SCHOOLS

JOHN D. SPENGLER, Harvard University, *Chair*
VIVIAN E. LOFTNESS, Carnegie Mellon University, *Vice Chair*
CHARLENE W. BAYER, Georgia Institute of Technology
JOHN S. BRADLEY, National Research Council, Ottawa, Canada
GLEN I. EARTHMAN, Virginia Polytechnic Institute and State University
PEYTON A. EGGLESTON, Johns Hopkins University
PAUL FISETTE, University of Massachusetts, Amherst
CAROLINE BREESE HALL, University of Rochester
GARY T. HENRY, University of North Carolina, Chapel Hill
CLIFFORD S. MITCHELL, Johns Hopkins University
MARK S. REA, Rensselaer Polytechnic Institute
HENRY SANOFF, North Carolina State University
CAROL H. WEISS, Harvard University (resigned September 2005)
SUZANNE M. WILSON, Michigan State University

Staff

LYNDA STANLEY, Director, Board on Infrastructure and the Constructed Environment
KEVIN LEWIS, Program Officer
PAT WILLIAMS, Senior Project Assistant

Acknowledgment of Reviewers

This report has been reviewed in draft form by individuals chosen for their diverse perspectives and technical expertise, in accordance with procedures approved by the National Research Council's Report Review Committee. The purpose of this independent review is to provide candid and critical comments that will assist the institution in making its published report as sound as possible and to ensure that the report meets institutional standards for objectivity, evidence, and responsiveness to the study charge. The review comments and draft manuscript remain confidential to protect the integrity of the deliberative process. We wish to thank the following individuals for their review of this report:

David W. Bearg, Life Energy Associates
Sheila Bosch, Green Ark, Inc.
Julie Dockrell, University of London
Dennis Dunne, dddunne & associates
James W. Guthrie, Vanderbilt University
Alan Hedge, Cornell University
James Kadamus, Sightlines, Inc.
Melvin Mark, Pennsylvania State University
Donald Milton, University of Massachusetts, Lowell
William B. Rose, University of Illinois at Urbana-Champaign
Ward V. Wells, Texas A&M University
Richard N. Wright, National Institute of Standards and Technology (retired)

Although the reviewers listed above have provided many constructive comments and suggestions, they were not asked to endorse the conclusions or recommendations, nor did they see the final draft of the report before its release. The review of this report was overseen by Henry W. Riecken, University of Pennsylvania, Emeritus. Appointed by the National Research Council, he was responsible for making certain that an independent examination of this report was carried out in accordance with institutional procedures and that all review comments were carefully considered. Responsibility for the final content of this report rests entirely with the authoring committee and the institution.

Contents

Executive Summary

School buildings are special places. They are the locus of education, the places where children come together to learn about civics and develop basic skills to be productive members of society. Schools are also used for adult education classes, voting, community events, and other activities, and may symbolize the community itself.

Research has shown that the quality of indoor environments can affect the health and development of children and adults. Buildings, including schools, also affect the natural environment, accounting for 40 percent of U.S. energy use and 40 percent of atmospheric emissions, including greenhouse gases. In the 1990s, the "green building" movement emerged to promote methods for designing buildings that have fewer adverse environmental impacts. Various guidelines have been developed to implement green building objectives.

SCOPE OF THE STUDY

At the request of the Massachusetts Technology Collaborative (MASSTECH), the Barr Foundation and the Kendall Foundation, the Connecticut Clean Energy Fund, and the U.S. Green Building Council, the National Research Council (NRC) appointed the Committee to Review and Assess the Health and Productivity Benefits of Green Schools. The committee's charge was to "review, assess, and synthesize the results of available studies on green schools and determine the theoretical and methodological basis for the effects of green schools on student learning and teacher productivity." The committee was also asked to look at the possible connections between green schools and student and teacher health. The results of this study should be of interest to a wide range of stakeholders, including school administrators, school district business managers, federal and state education officials, parents, and teachers, as well as architects and engineers specializing in school design, both green and conventional.

COMPLEXITY OF THE TASK AND THE COMMITTEE'S APPROACH

The line of reasoning inherent in this study's task—mapping connections from physical environments to student and teacher outcomes (health, learning, productivity)—poses significant challenges. Numerous factors contribute to the complexity of the task, including the lack of a clear definition of what constitutes a green school, the variations in current green school guidelines, the difficulty of measuring educational and productivity outcomes, the variability and quality of the scientific literature, and confounding factors. All of these factors are described in Chapter 2.

Lacking specific guidance, the committee identified those building characteristics and practices typically emphasized in current green school guidelines. The committee determined that green schools have two complementary, but not identical, goals and resulting outcomes. The goals are (1) to support the health and development (physical, social, intellectual) of students, teachers, and staff by providing a healthy, safe, comfortable, and functional physical environment; and (2) to have positive environmental and community attributes. Because they were first developed to minimize adverse environmental effects, current green school guidelines place less emphasis on features supporting human health and development. In line with its charge, the committee focused on outcomes associated with student and teacher health, learning, and productivity.

The committee developed a conceptual model for evaluating the links between green school buildings and outcomes (Figure ES.1).

The conceptual model assumes that a green school building's location and design (site; orientation; envelope; heating, ventilation, and air conditioning; acoustics; lighting) will result in an indoor environment with appropriate (or inappropriate) levels of moisture, ventilation, air quality, noise, lighting, and other qualities. It also assumes that the indoor environment will be modified by season (e.g., presence of airborne pollen), over time (e.g., mold growth from chronic water leakage), and by operational, maintenance, repair, and cleaning practices. Finally, the indoor environment can affect student learning and health and teacher health and productivity.

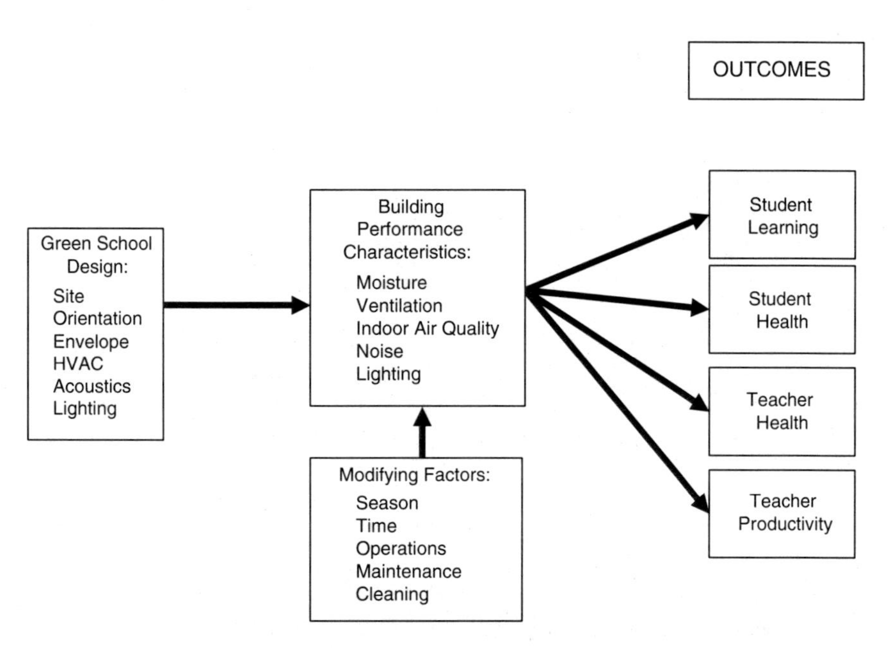

FIGURE ES.1 Conceptual model for evaluating links between green school buildings and outcomes for learning, health, and productivity.

GREEN SCHOOL BUILDING ATTRIBUTES THAT WOULD SUPPORT HEALTH AND DEVELOPMENT

After evaluating the research literature, the committee concluded that a green school with the following attributes would support student and teacher health, learning, and productivity:

- *Dryness:* Excessive moisture, which has been associated with adverse health effects, particularly asthma and respiratory diseases, is not present.
- *Good indoor air quality and thermal comfort:* Ventilation rates, air pollutants, humidity levels, and temperature ranges, which have been linked to human health, learning, and productivity, are effectively controlled.
- *Quietness:* The acoustical quality, which has been shown to affect student learning and the development of language skills, meets the newly released Standard 12.60, "Acoustical Performance Criteria, Design Requirements, and Guidelines for Schools," of the American National Standards Institute.

- *Well-maintained systems:* Building systems are commissioned[1] to ensure that they perform as intended, and their performance is monitored over time. Routine preventive maintenance is implemented throughout a school's service life.
- *Cleanliness:* Surfaces are disinfected to interrupt the transmission of infectious diseases, and measures are implemented to help control indoor pollutants that have been associated with asthma and other respiratory diseases.

Additional green school attributes that should be aspired to include durability, increased acoustical quality for more sensitive groups, and improved cleaning practices to prevent the accumulation of allergens and irritants. The committee's specific findings and recommendations are presented below.

FINDINGS AND RECOMMENDATIONS

Finding 1: School buildings are composed of many interrelated systems. A school building's overall performance is a function of interactions among these systems, of interactions with building occupants, and of operations and maintenance practices. However, school buildings are typically designed by specifying individual components or subsystems, an approach that may not recognize such interactions. Current green school guidelines reflect this approach, and in so doing, they allow for buildings focused on specific objectives (e.g., energy efficiency) at the expense of overall building performance.

Recommendation 1a: Future green school guidelines should place greater emphasis on building systems, their interrelationships, and overall performance. Where possible, future guidelines should identify potential interactions between building systems, occupants, and operation and maintenance practices and identify conflicts that will necessitate trade-offs among building features to meet differing objectives.

Recommendation 1b: Future green school guidelines should place greater emphasis on operations and maintenance practices over the lifetime of a building. Systems that are durable, robust, and easily installed, operated, and maintained should be encouraged.

[1]Commissioning is "a quality-focused process for enhancing the delivery of a project. The process focuses on verifying and documenting that the facility and all of its systems and assemblies are planned, designed, installed, tested, operated to meet the Owner's Project Requirements" (ANSI, 2004, p. 1). Commissioning is discussed in detail in Chapter 9.

Complexity of the Task

Finding 2a: Given the complexity of interactions between people and their environments, establishing cause-and-effect relationships between an attribute of a green school or other building and its effect on people is very difficult. The effects of the built environment may appear to be small given the large number of variables and confounding factors involved.

Finding 2b: The committee did not identify any well-designed, evidence-based studies concerning the *overall* effects of green schools on human health, learning, or productivity or any evidence-based studies that analyze whether green schools are actually different from conventional schools in regard to these outcomes. This is understandable because the concept of green schools is relatively new, and evidence-based studies require a significant commitment of resources.

Finding 2c: Scientific research related to the effects of green schools on children and adults will be difficult to conduct until the physical characteristics that differentiate green from conventional schools are clearly specified. With outcomes as complex as student and teacher health, student learning, and teacher productivity influenced by many individual family and community factors, it may be possible in theory to design research that controls for all potentially confounding factors, but difficult in practice to conduct such research.

Recommendation 2: The attributes of a green school that may potentially affect student and teacher health, student learning, and teacher productivity differently than those in conventional schools should be clearly specified. Once specified, it may be possible to design appropriate research studies to determine whether and how these attributes affect human health, learning, and productivity.

Building Envelope, Moisture Management, and Health

Finding 3a: There is sufficient scientific evidence to establish an association between excess moisture, dampness, and mold in buildings and adverse health outcomes, particularly asthma and respiratory symptoms, among children and adults.

Finding 3b: Excess moisture in buildings can lead to structural damage, degraded performance of building systems and components, and cosmetic damage, all of which may result in increased maintenance and repair costs.

Finding 3c: Well-designed, -constructed, and -maintained building envelopes are critical to the control and prevention of excess moisture and molds. Designing for effective moisture management may also have benefits for the building, such as lower life-cycle costs.

Finding 3d: Current green school guidelines typically do not adequately address the design detailing, construction, and long-term maintenance of buildings to ensure that excess moisture is controlled and a building is kept dry during its service life.

Recommendation 3a: Future green school guidelines should emphasize the control of excess moisture, dampness, and mold to protect the health of children and adults in schools and to protect a building's structural integrity. Such guidelines should specifically address moisture control as it relates to the design, construction, operation, and maintenance of a school building's envelope (foundations, walls, windows, and roofs) and related items such as siting and landscaping.

Recommendation 3b: Research should be conducted on the moisture resistance and durability of materials used in school construction. Such research should also investigate other properties of these materials such as the generation of bioaerosols and indoor pollutants as well as the environmental impacts of producing and disposing of these materials.

Indoor Air Quality, Health, and Performance

Finding 4a: A robust body of scientific evidence indicates that the health of children and adults can be affected by indoor air quality. A growing body of evidence suggests that teacher productivity and student learning may also be affected by indoor air quality.

Finding 4b: Key factors in providing good indoor air quality are the ventilation rate; ventilation effectiveness; filter efficiency; the control of temperature, humidity, and excess moisture; and operations, maintenance, and cleaning practices.

Finding 4c: Indoor air pollutants and allergens from mold, pet dander, cockroaches, and rodents also contribute to increased respiratory and asthma symptoms among children and adults. Although limited data are available regarding exposure to these allergens in U.S. schools, studies in both school and nonschool environments support the notion that allergen levels can be decreased through good cleaning practices.

Finding 4d: The reduction of pollutant loads through increased ventilation and effective filtration has been shown to reduce the occurrence of building-associated symptoms (eye, nose, and throat irritations; headaches; fatigue; difficulty breathing; itching; and dry, irritated skin) and to improve the health and comfort of building occupants.

Finding 4e: There is evidence that ventilation rates in many schools do not meet current standards of the American Society for Heating, Refrigeration, and Air-Conditioning Engineers (ASHRAE). Available research indicates that increasing the ventilation rate to exceed the current ASHRAE standard will further improve comfort and productivity.

Finding 4f: Scientific evidence indicates that increased ventilation rates can reduce the incidence of building-related symptoms, reduce pollutant loads associated with asthma and other respiratory diseases, and improve the productivity of adult workers. Increased ventilation rates may also reduce the potential for reactions among airborne pollutants that generate irritating products and may improve perceived air quality. However, the research conducted to date has not established an upper limit on the ventilation rates, above which the benefits of outside air begin to decline.

Finding 4g: Research comparing the effects of natural versus mechanical ventilation on human health is inconclusive. However, there is evidence that improper design, maintenance, and operation of mechanical ventilation systems contribute to adverse health effects, including building-related symptoms among occupants.

Finding 4h: Studies in office buildings indicate that productivity declines if room temperatures are too high. However, there are few studies investigating the relationship of room temperatures to student learning, teacher productivity, and occupant comfort.

Finding 4i: To date, no systematic research has examined the relationship of cleaning effectiveness to student and teacher health, student learning, or teacher productivity. Few studies have looked systematically at changes in exposures, health, or productivity based on changes in building materials, cleaning products, or cleaning practices.

Recommendation 4a: Future green school guidelines should ensure that, as a minimum, ventilation rates in schools meet current ASHRAE standards overall and as they relate to specific spaces. Future guidelines should also give consideration to planning for ventilation systems that can be easily adapted to meet evolving standards for ventilation rates, temperature, and humidity control.

Recommendation 4b: Future green school guidelines should emphasize the importance of appropriate operation and preventive maintenance practices for ventilation systems, including replacing filters, cleaning coils and drip pans to prevent them from becoming a source of air pollution, microbial contamination, and mold growth. These systems should be designed to allow easy access for maintenance and repair. The Environmental Protection Agency's Tools for Schools program is a well-recognized source of information on methods for achieving good indoor air quality.

Recommendation 4c: Additional research should be conducted to document the full range of costs and benefits of ventilation rates that exceed the current ASHRAE standard and to determine optimum temperature ranges for supporting student learning, teacher productivity, and occupant comfort in school buildings.

Recommendation 4d: Studies should be conducted to examine the relationships of exposures from building materials, cleaning products, and cleaning effectiveness to student and teacher health, student learning, and teacher productivity.

Lighting and Human Performance

Finding 5a: The research findings from studies of adult populations seem to indicate clearly that the visual conditions in schools resulting from both electric lighting and natural light (daylighting) should be adequate for most children and adults, although this supposition cannot be supported by direct evidence.

Finding 5b: There is concern that a significant percentage of students in classrooms do not have properly corrected eyesight, and so the general lighting conditions suitable for visual functioning by the most students may be inadequate for those students who need but do not have corrective lenses. It could be hypothesized that daylight might benefit these children by providing higher light levels and better light distribution (side light) than would electric lighting alone. However, the potential advantages of daylight in classrooms for improving the visual performance of children without properly corrected eyesight has not been systematically studied.

Finding 5c: Current green school guidelines typically focus on energy-efficient lighting technologies and components and the use of daylight to further conserve energy when addressing lighting requirements. Guidance for lighting design that supports the visual performance of children and adults, based on task, school room configurations, layout, and surface finishes, is not provided.

Finding 5d: Windows and clerestories can supplement electric light sources, providing high light levels, and good color rendering. Light from these sources is ever-changing and can cause glare unless appropriately managed. Currently, there is insufficient scientific evidence to determine whether or not an association exists between daylight and student achievement.

Finding 5e: A growing body of evidence suggests that lighting may play an important nonvisual role in human health and well-being through the circadian system. However, little is known about the effects of lighting in schools on student achievement or health through the circadian system.

Recommendation 5a: Future green school guidelines should seek to support the visual performance of students, teachers, and other adults by encouraging the design of lighting systems based on task, school room configurations, layout, and surface finishes. Lighting system performance should be evaluated in its entirety, not solely on the source of illumination or on individual components.

Recommendation 5b: Future green school guidelines for the design and application of electric lighting systems should conform to the latest published engineering practices, such as the consensus lighting recommendations of the Illuminating Engineering Society of North America.

Recommendation 5c: Green school guidelines that encourage the extensive use of daylight should address electric control systems and specify easily operated manual blinds or other types of window treatments to control excessive sunlight or glare.

Recommendation 5d: Because light is important in regulating daily biological cycles, both acute effects on learning and lifelong effects on children's health should be researched, particularly the role that lighting in school environments plays in regulating sleep and wakefulness in children.

Acoustical Quality, Student Learning, and Teacher Health

Finding 6a: Most learning activities in school classrooms involve speaking and listening as the primary communication modes. The intelligibility of speech in classrooms is related to the levels of speech sounds relative to the levels of ambient noise and to the amount of reverberation in a room.

Finding 6b: Sufficient scientific evidence exists to conclude that there is an inverse association between excessive noise levels in schools and student learning.

Finding 6c: The impacts of excessive noise vary according to the age of students, because the ability to focus on speech sounds is a developmental skill that does not mature until about the ages of 13 to 15 years. Thus, younger children require quieter and less reverberant conditions than do adults to hear equally well. As adults, teachers may not appreciate the additional problems that excessive noise creates for younger students.

Finding 6d: Excessive noise is typically a more significant problem than is too much reverberation in a classroom. It is not possible to have both increased speech level (to maximize signal to noise) and reduced reverberation times. Good acoustical design must be a compromise that strives to increase speech levels without introducing excessive reverberation.

Finding 6e: The most substantial body of research related to excessive noise and learning in the classroom addresses the impacts of road traffic, trains, and airport noise.

Finding 6f: Some available evidence indicates that teachers may be subject to voice impairment as a result of prolonged talking in noisy school environments. However, there is no information to quantify a relationship between specific noise levels in classrooms and potential voice impairment.

Recommendation 6a: To facilitate student learning, future green school guidelines should require that new schools be located away from areas of higher outdoor noise such as that from aircraft, trains, and road traffic.

Recommendation 6b: Future green school guidelines should specify acceptable acoustical conditions for classrooms and should require the appropriate design of HVAC systems, the design of walls and doors separating classrooms and corridors, and the acoustic quality of windows and walls adjoining the outdoors. This recommendation is most easily achieved by requiring that green schools comply with American National Standards Institute (ANSI) Standard 12.60, "Acoustical Performance Criteria, Design Requirements, and Guidelines for Schools."

Recommendation 6c: Additional research should be conducted to define optimum classroom reverberation times more precisely for children of various ages.

Building Characteristics and the Spread of Infectious Disease

Finding 7a: Common viruses and infectious diseases can be transmitted by multiple routes: through the air, by person-to-person contact, and by touching contaminated surfaces (fomites). Certain characteristics of buildings, including the cleanliness of surfaces, relative humidity, and ventilation effectiveness, influence the transmission of common viruses. Evidence from studies in nonschool environments suggests that interventions which interrupt the known modes of transmission of common infectious agents may decrease the occurrence of such illnesses in schoolchildren and staff.

Finding 7b: The best way to control infections, especially gastroenteritis, appears to be instituting procedures that promote good hand cleansing. Available but limited information indicates that hand sanitizers are superior to the routine washing of hands.

Finding 7c: Cleaning of surfaces that are commonly touched (e.g., doors, faucets, desktops) is effective for interrupting the transmission of infectious agents. Disinfecting surfaces with water and detergents is apparently as effective as applying germicidal agents.

Finding 7d: The use of no-touch faucets, doorways, receptacles, and equipment seems to be a reasonable, though unproven, method for infection control.

Finding 7e: The survival, dispersal, and removal of airborne pathogens are affected by relative humidity, ventilation rate, and the percentage of recirculated air in the air supply. Increased ventilation rates have been shown to speed the dilution and removal of viral material. The use of displacement ventilation and the reduction of the percentage of recirculated air in the air supply have the potential to reduce building occupants' exposures to airborne pathogens.

Finding 7f: Ultraviolet germicidal irradiation (UVGI) may be effective for inactivating and killing some infectious organisms, but its use in school room applications has not been systematically studied.

Recommendation 7a: Future green school guidelines should include measures for the regular cleaning of commonly touched surfaces and the availability of hand sanitizers at sinks. The use of "no-touch" faucets, receptacles, equipment, and egress from bathrooms should be considered, taking into account the age of the children in the school.

Recommendation 7b: Full-scale classroom and school studies should be conducted to quantify the efficacy of a variety of ventilation strategies, including displacement ventilation and the elimination of recirculated air, for the dispersion and removal of airborne infectious agents. Studies should also quantify the potential costs and benefits of such ventilation strategies.

Recommendation 7c: Additional research should be conducted to determine the optimal infection-control interventions in terms of measurable outcomes such as absenteeism and academic achievement. One line of research is the use of ultraviolet germicidal irradiation in supplemental or portable air-cleaning devices in school room applications and its effects on human health.

Overall Building Condition and Student Achievement

Finding 8: The methodologies used in studies correlating *overall* building condition with student achievement are not adequate to determine if there is a relationship between *overall* building condition and student test scores. This research tradition seems to address a more general and diffuse question and does not produce high-quality evidence relative to either school design or specific aspects of maintenance. Improved research for understanding how specific building conditions affect student and teacher performance would measure one or more building performance characteristics, develop a theory linking those characteristics and student and/or teacher outcomes, and test the linkage using adequate measures of the outcomes of interest and fully specified regression models.

Processes and Practices for Planning and Maintaining Green Schools

Finding 9a: Participatory planning, commissioning, and postoccupancy evaluation are processes that can both lower building operating costs and improve performance over a building's lifetime. Current green school guidelines typically only address the practice of building commissioning.

Finding 9b: Inadequate planning for schools carries long-term fiscal, human, and academic costs. A strong planning process requires asking the right questions, involving a full range of stakeholders, and having a clear sense of purpose.

Finding 9c: A commissioning process that starts in the planning phase and continues through building occupancy can help ensure that a school building performs in accordance with the stated design criteria and the owner's operational requirements. Effective commissioning for green schools requires specific expertise in nontraditional elements such as moisture control, indoor air quality, lighting, and acoustics.

Finding 9d: If a green school's performance and potential benefits are to be maintained over its service life, building systems and features should be monitored. Such monitoring can include the use of sensors and other technologies that provide data about current indoor environmental conditions and the likely performance of a building over time.

Finding 9e: Postoccupancy evaluations can help ensure the performance of existing schools and help improve the design of future schools.

Finding 9f: Green schools represent a significant public investment. That investment can be undermined if educators, support staff, students, and other stakeholders do not have the knowledge or training to appropriately use or operate a green school.

Recommendation 9a: Future green school guidelines should stress the importance of good planning processes that allow for the effective participation of a wide range of stakeholders.

Recommendation 9b: Future green school guidelines should require for all new schools a building commissioning process that begins in planning and continues through occupancy. The commissioning agent should specifically verify that moisture-management features are properly designed and installed, that intended ventilation rates are delivered to building occupants, that the lighting system is adequately designed and installed to ensure effective lighting based on tasks and school room configurations, and that acoustical measures meet the performance standards of ANSI Standard 12.60.

Recommendation 9c: Future green school guidelines should encourage the periodic monitoring of indoor environmental characteristics including moisture levels, absolute humidity, classroom temperatures, and ventilation effectiveness to ensure that performance objectives are maintained over the service life of a school.

Recommendation 9d: Educators, support staff, students, and other stakeholders should be informed of the design intent of a green school and given the appropriate information or training to fulfill their roles in using and operating a green school.

Linking Green Schools to Health and Productivity: Research Considerations

Finding 10a: Much is still not known about the potential interactions of building systems, materials, operation and maintenance practices and their effects on building occupants in general, or about school environments in particular. The necessary collaboration between architecture, engineering, physical science, medicine, and social science expertise is a challenge, but multidisciplinary research is required to fully study the potential relationship between a school building and the outcomes of students and teachers.

Finding 10b: In designing research studies to evaluate the unbiased effects of green schools on student learning or student and teacher health, several issues must be addressed. These include defining green schools for the purpose of scientific inquiry, defining performance and productivity outcomes plausibly related to green schools, and fully developing a theory explaining the links between green school design and health and learning effects. Finally, the hypotheses from these theories should be tested in ways that reduce systematic biases and provide compelling evidence about these linkages.

Finding 10c: Currently, the theory and evidence connecting green schools or characteristics associated with green schools to teacher or student outcomes is not sufficient to justify large-scale evaluations. However, the committee does consider it useful to carry out studies that assess the positive and negative consequences of the design and construction features as well as building performance characteristics that are associated with green schools using more rigorous study designs.

Finding 10d: Large-scale evaluations using randomized experiments and econometric or regression-based techniques should be conducted if they are justified from the results of smaller and less expensive studies, such as those outlined in Finding 10c. Finally, it is possible that improvements to the large-scale data sets that contain student achievement data will allow for relatively low cost studies of the effects of the school building environment on student achievement, which the committee considers to be an important side benefit.

1

Introduction

Americans have long cherished the belief that a well-educated citizenry is necessary to the national well-being. School buildings are the places where children come together to learn basic civics as well as the skills they need to become productive members of society: Schools are the locus of education. Local schools have been the center of efforts to provide equal educational opportunities to all segments of the population. School buildings also host other activities, including adult education classes, voting, and community events. In some cases, a school may symbolize the community itself and may have intrinsic value for social coherence.

The concept that the design of school buildings may affect students' and teachers' health and development is not new. A report on the State of Maine's Schools in 1886 linked moisture, lighting, and ventilation of school buildings to health and learning:

> Nearly one sixth of the population of our State spends about six hours daily during a large part of the year in our school rooms. This necessary confinement within the school room walls, coming as it does during the growing period of the body, and while it is most susceptible to harmful influences, entails certain evils which have been too generally regarded as necessary accompaniments of school life. It is generally well known, however. . . . that most of the diseases incident to school life are in quite a high degree preventable. . . . [O]ne of the first and most important requirements in guarding against these diseases is to have the building of school-houses conform to a few rules. . . .

- School-houses should be built on dry ground; if not dry, the lots must be deeply drained. . . . The reason for this rule is the well-known fact that dampness of soil contributes much to make a school-house unhealthy. . . .
- The light which comes from considerably above the level of the desks and books lights them much better than the more horizontal rays. High windows also light the ceiling, whence the light is reflected downward upon the desks. . . .
- Never think that a school-room is completed until there is some way of getting fresh, warmed air into it and the foul, breathed air out. Ventilation costs something in fuel, but it is a penny-wise and pound-foolish policy which omits it (State of Maine, 1887).

More than a hundred years later, scientific research has demonstrated that building design, materials, systems, operation, maintenance, and cleaning practices can affect occupants' health and development.

Buildings also affect the natural environment, in the resources used and the pollutants emitted. As shown in Table 1.1, buildings account for more than 40 percent of U.S. energy use as well as a significant amount of raw materials, water, and land. Buildings also produce 40 percent of atmospheric emissions, including greenhouse gases, and significant amounts of solid waste and wastewater.

Recognizing these adverse environmental impacts, a movement to design and operate buildings using methods and technologies that conserve energy and other natural resources—often called green building, high-performance building, or sustainable design—emerged in the 1990s and continues to grow. This movement emphasizes designing, constructing, operating, and maintaining buildings to reduce their adverse environmental impacts through the use of recycled materials, energy-efficient equipment, and other features and practices. The potential environmental benefits of green school buildings are significant: A 2002 survey of 851 public schools districts found that an average of $176 per pupil was spent for energy. This figure is likely to be higher in 2006 owing to across-the-board increases in the price of gas, oil, and electricity. If energy and other

TABLE 1.1 Environmental Burdens of Buildings, U.S. Data

Resource Use	Share of Total (%)	Pollution Emissions	Share of Total (%)
Raw materials	30	Atmospheric emissions	40
Energy use	42	Water effluents	20
Water use	25	Solid waste	25
Land (SMSAs[a])	12	Other releases	13

[a]Standard Metropolitan Statistical Areas.
SOURCE: Data from Levin (1997).

building-related costs can be reduced, the money saved could be used for other educational purposes.

Because research on the effects of the indoor environment on people and research on the effects of buildings on the environment are converging, there is value in determining whether some building designs, technologies, and practices that help to support human health and development can also benefit the natural environment. This study, then, is intended to look beyond the environmental objectives established for "green" schools and to assess the research-based evidence related to their effects on student learning, teacher productivity, and the health of students, teachers, and staff. The results of this study should be of interest to a wide range of stakeholders, including school administrators, school district business managers, federal and state education officials, parents, teachers, and architects and engineers specializing in school design, both green and conventional.

SCHOOL CONSTRUCTION AND RELATED ISSUES

More than 62 million Americans spend a significant portion of their time in school buildings. The National Center for Education Statistics (NCES) reports that in 2002 more than 123,000 elementary and secondary schools, with a combined enrollment of 54 million students, were operating in the United States (Table 1.2) (NCES, 2002). Approximately 7.3 million people were employed to provide teaching and education-related services in public schools: 3 million teachers; 2.9 million education-related staff; and 1.4 million support services staff. Complete data for private schools were not available. On average, public schools invested $7,900 per pupil per year to provide educational services, which equates to $39.5 million per year for a public school district with 5,000 students.

TABLE 1.2 Statistics for Elementary and Secondary Schools in the United States, 2002

	Total	Public	Private
Schools	123,382	94,112	29,270
Students enrolled (in millions)	54.6	48.2	6.4
Teachers employed (FTEs)	3,425,400	3 million	425,400
Education-related staff		2.9 million	Not available
Support services staff[a]		1.4 million	Not available
Average expenditure per student[b]		$7,900	

[a]Includes school and district administrators, librarians, guidance counselors, instructional aides, and supervisors.
[b]Includes persons providing food, health, library assistance, maintenance, transportation, security, and other services.
SOURCE: NCES (2002).

The NCES projects that by 2014, total enrollments will increase by approximately 2 million, to 56.7 million, and the number of teachers employed will increase by 500,000, to 3.9 million. Expenditures per public school student are forecasted to increase to $10,000 by 2014, or $50 million for a district with 5,000 public school students.

The primary responsibility for providing an educational system rests with each of the 50 states and the District of Columbia. The majority of funding for educational programs is provided by state and local authorities, although the federal government does provide some funding tied to policy initiatives such as the No Child Left Behind Act (P.L. 107-110).

School building construction and renovation also require a substantial commitment of resources—dollars, time, materials, expertise. Capital costs alone typically will total tens of millions of dollars. The decision to invest this level of funding most often rests with the local school governing body and usually requires a bond referendum voted upon by the residents of the school district. The process of authorizing funding for construction may take several years (Figure 1.1). Once funds are allocated, it may take 2 to 5 more years for a school building to be designed, constructed, and occupied.

A school building's size and design (which affect initial capital costs and long-range operating costs) are a function of the current and projected school population, local resources, the curriculum, and other community uses. For example, the educational literature abounds with articles touting the virtues of small neighborhood schools (Cotton, 2001; Raywid, 1996). Advocates for smaller schools argue that such schools are better at improving the academic achievement of students who have not been successful in traditional settings, leading to higher graduation rates, and other benefits. Others argue that the cost of moving to smaller schools is too great. To date, the scientific evidence is mixed as to whether smaller or larger schools produce better academic results.

The ongoing debate about the purpose and nature of public education also has implications for school building design, and costs. If public education becomes increasingly focused on producing good scores on standardized achievement tests, for instance, curriculums may increasingly focus on traditional academic subjects, and the demand for music, art, vocational, and physical education courses may diminish. School buildings designed to support this type of curriculum might be composed primarily of standard academic classrooms with few spaces for so-called "nonessential" subjects. In contrast, a curriculum based on the premise that learning is holistic—with, for example, art incorporated into language arts or math taught with specific job skills or vocations in mind—may require school buildings in which standard academic class-

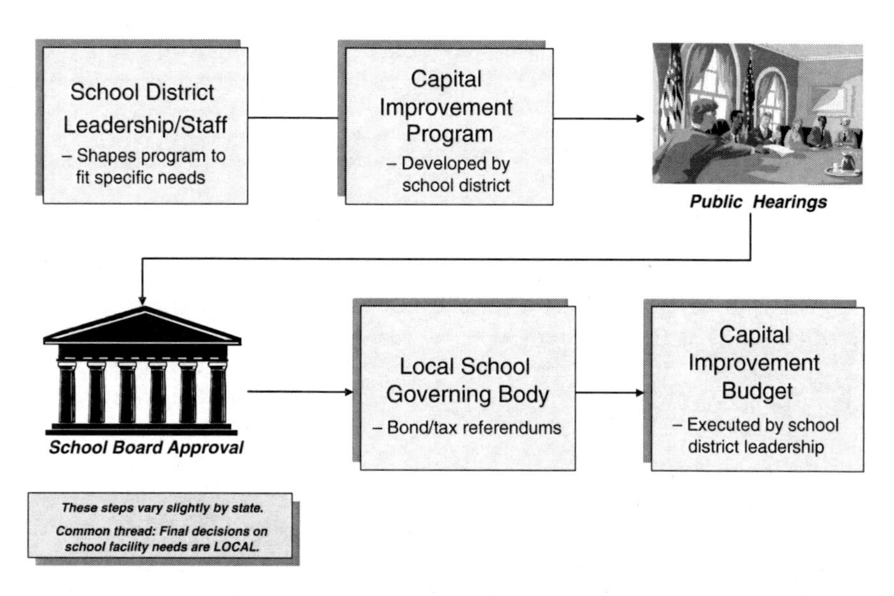

FIGURE 1.1 Typical process for funding school buildings.

rooms largely disappear, to be replaced by specialized labs and learning centers (Lackney, 1999). Another possibility is schools that are created or redesigned so that instructional and support spaces can also be used by social and community organizations or even businesses. Schools in which teaching might be individualized through Web-based teaching are also being discussed. Classrooms in such schools would contain computers but few other materials, such as books. Thus, public discussion and decision making about the size of schools—large or small—and educational curriculums, have significant implications for the design of schools, construction materials and resources, their long-term operation and maintenance costs, and the tax dollars that will be invested in them. Implicit in these discussions is the assumption that school design does have a role in supporting educational goals and outcomes.

Once built, a school typically is used for educational purposes for 30 years or longer. During that time, changes in functional requirements or new programmatic ideas may drive the reconfiguration of classrooms or the upgrading of libraries, science laboratories, or other specialized space.

While the building is in service, the investment made in its operation, maintenance, and repair will be six to eight times greater than the initial cost of construction.[1] For that reason, a focus on the lifetime costs of a building, not just the first costs of design and construction, is important for effective decision making and the long-term economic health of a community.

A SCHOOL BUILDING AS A SYSTEM OF SYSTEMS

A school, like other buildings, is a system of systems. A school building's overall performance is a function of the indoor and outdoor environments, the activities taking place within it and on its grounds, the occupants, how intensely it is used (9 or 12 months of the year, by how many people), the initial design, and long-term operation and maintenance practices.

Indoor environmental quality—the level of air pollutants, temperature, humidity, noise, light, space—is a function of the interrelationship of a building's foundations, floors, walls, roofs, heating, ventilation, and air conditioning (HVAC) systems, electrical and plumbing systems, telecommunications, materials, and furnishings. HVAC systems, for example, affect a range of indoor environmental factors, including pollutant levels, temperature, humidity, noise, air quality, moisture control, and sensory loads (odors). The location of air intakes, the efficiency of ventilation filters, and operation practices all will affect the amount and quality of outdoor air used to ventilate indoor spaces. The optimum size and capacity of an HVAC system is dependent on the orientation of the building, the total square footage, the quality of insulation, the number of windows, and other factors. The orientation of a school to maximize solar gain might help reduce heating costs during the winter but increase cooling costs in the summer. If the HVAC system design does not adequately factor in these variables, the system may be too large and energy inefficient.

Numerous other examples of the interrelationships between the design and operation of a building system and the indoor environmental quality could be cited. In all cases, indoor environmental quality can and will deteriorate over time if buildings are not properly designed, systems are not operated appropriately, or needed maintenance and repairs are deferred.

Although the interactions and interdependencies of building systems and their various components are known by building designers, individual systems and elements are typically treated as separate components,

[1]The annual operation and maintenance costs of a building, however, will be only a fraction of the annual costs to operate a school, which includes the salaries and benefits of teachers, administrators, and support staff; educational equipment and supplies; food service; and other expenses.

with attention devoted to the specification and performance of each individual structural subsystem. This sequential approach to design and construction reflects the organization of the design and construction industry into individual trades and specialties. Research on the effects of the built environment on people also tends to focus on specific building features or practices as opposed to overall design or building performance.

Because building systems and components are designed and installed as independent entities, some systems may be optimized at the expense of others, and potential conflicts among design objectives may not be recognized. For example, providing for good acoustical quality in classrooms may require that ventilation and air-conditioning ducts be lined with sound-absorbing materials. However, if some materials used for this purpose are not properly installed and maintained, microbial contamination can occur and negatively affect the indoor air quality. A "systems" philosophy of design considers these types of interactions and recognizes that compromises or trade-offs among elements may be necessary to optimize overall building performance.

When school building design and performance are being considered, the element of time cannot be overemphasized. School building materials and components wear out at differing rates, vary in their complexity, the ways in which they are operated, and the costs of maintaining them. So, although a building's foundations and walls may last for 50 to 100 years, the roof will probably wear out after 20 years and the air-conditioning system in 15 years. The durability of materials, the level of maintenance undertaken, the timeliness and quality of the maintenance, the climate, and other factors will affect the service lives and performance of various systems and components.

SCHOOL BUILDING PERFORMANCE

Although most school buildings perform well when first built, their performance can and will deteriorate if the systems are not operated appropriately, if preventive maintenance programs are ineffective, or if needed maintenance and repairs are deferred. Because overall building performance is difficult to measure directly, overall building condition is often used as a surrogate. Professional organizations and governmental agencies have been reporting on the condition of the nation's schools (AASA, 1983; Council of Great City Schools, 1987; Educational Writers' Association, 1989; GAO, 1995a, 1995b; NEA, 2000) for 25 years. These reports consistently found that a substantial portion of the school-age population was being educated in substandard buildings. And schools with higher concentrations of students from low-income households were more likely to be in substandard condition.

In *School Facilities: Condition of America's Schools*, the U.S. General Accounting Office estimated that approximately 14 million students (30 percent of all students) attended schools that needed extensive repairs or replacement (one-third of the school inventory) as of 1995 (GAO, 1995a). Approximately 28 million students attended schools that needed extensive repairs on one or more major building systems. The building components or features most often identified as needing attention in substandard schools were thermal control (temperature and humidity), ventilation, plumbing, roofs, exterior walls, finishes, windows, doors, electrical power, electrical lighting, life safety (fire suppression), and interior finishes and trims. The cost to make necessary repairs was estimated at more than $100 billion (1995 dollars).

A second GAO report, *School Facilities: America's Schools Not Designed or Equipped for 21st Century*, found that approximately 40 percent of the schools surveyed could not meet the functional requirements for teaching laboratory science or large groups (GAO, 1995b). About two-thirds of the schools could not support educational reform measures such as private space for counseling and testing, parental support activities, social/health care, day care, and before- and after-school care.

In 2000, the NCES reported that at least 29 percent of the nation's public elementary and secondary schools had problems with heating, ventilation, and air conditioning; 25 percent had plumbing problems; 24 percent reported problems with exterior walls, finishes, windows, and doors; and about 20 percent had less than adequate life safety, roofs, and electrical power. About 11 million students attended school in districts reporting less-than-adequate buildings, of whom approximately 3.5 million were in schools whose condition was rated as poor, which needed to be replaced, or in which significant substandard performance was apparent (NCES, 2000). The NCES also found that:

- Schools in rural areas and small towns were more likely than schools in urban fringe areas and large towns to report that at least one of their environmental conditions was unsatisfactory (47 percent compared with 37 percent),
- Schools with the highest concentrations of households with incomes below the poverty level were more likely to report at least one unsatisfactory environmental condition than were schools with lower concentrations of low-income households (55 percent compared with 38 percent), and
- About one-third of school administrators were dissatisfied with the energy efficiency of their schools, and 38 percent were dissatisfied with the flexibility of instructional space.

"GREEN" BUILDING MOVEMENT

Although the history of environmentally responsive design is centuries old, the term "green building" (also called "sustainable design building" or "high-performance building") is relatively new. The goal is to design buildings that meet performance objectives for land use, transportation, energy efficiency, indoor environmental quality, and other factors.

Statements describing green objectives vary. The Office of the Federal Environmental Executive, for example, defines green building as "the practice of (1) increasing the efficiency with which buildings and their sites use energy, water, and materials, and (2) reducing building impacts on human health and the environment, through better siting, design, construction, operation, maintenance, and removal—the complete building life cycle" (OFEE, 2003). An Urban Land Institute publication (Porter, 2000, p. 12) defines "sustainable" as designing projects and buildings in ways that

- Conserve energy and natural resources and protect air and water quality by minimizing the consumption of land, the use of other nonrenewable resources, and the production of waste, toxic emissions, and pollution;
- Make cost-effective use of existing and renewable resources such as infrastructure systems, underused sites, and historic neighborhoods and structures;
- Contribute to community identity, livability, social interaction, and sense of place;
- Widen access to jobs, affordable housing, transportation choices, and recreational facilities; and
- Expand diversity, synergism, and use of renewable resources in the operation and output of the local economy.

To enable building owners and design teams to evaluate the energy and environmental performance of their buildings, a voluntary system for rating green buildings has been developed for new construction by an expert-intensive process of the U.S. Green Building Council (USGBC). The USGBC's Leadership in Energy and Environmental Design (LEED) rating system awards credits for designing to advance sustainable site development, water savings, energy efficiency, indoor environmental quality, and materials selection. Buildings are rated as "certified," "silver," "gold," or "platinum," depending on the total number of credits received. To receive LEED certification a building must first comply with the most up-to-date national standards. Credits can also be earned for innovative designs.

The first version of LEED was issued in 1998 and has since been revised several times (the most recent version is LEED 2.2). The revisions have involved hundreds of environmental professionals and industries engaged in defining energy and environmental sustainability for a range of building types, for commercial interiors (LEED-CI), for existing buildings (LEED-EB), for commercial construction and major renovation projects (LEED-NC), and for neighborhood development (LEED-ND).

GREEN SCHOOL GUIDELINES

A consortium of state and utility leaders in California launched an effort in 2001 to develop energy and environmental standards specifically for schools. The Collaborative for High Performance Schools (CHPS, often pronounced "chips") aims to increase the energy efficiency of California schools by marketing information, services, and incentive programs directly to school districts and designers. The CHPS Web site defines green schools as having the following 13 attributes: "healthy, comfortable, energy efficient, material efficient, water efficient, easy to maintain and operate, commissioned, environmentally responsive site, a building that teaches, safe and secure, community resource, stimulating architecture, and adaptable to changing needs" (CHPS, 2005). Green school objectives are to be achieved through guidelines that are similar to the LEED rating system but specifically geared to schools. Similar guidelines have been issued by Washington state (WSBE, 2005) and are in development in Massachusetts and other states.[2] Green school guidelines move well beyond design and engineering criteria for the buildings themselves, addressing land use, processes for construction and equipment installation, and operation and maintenance practices. They include design and engineering techniques to meet specific objectives:

- Locating schools near public transportation to reduce pollution and land development impacts;
- Placing a building on a site so as to minimize its environmental impact and make the most of available natural light and solar gain;
- Designing irrigation systems and indoor plumbing systems to conserve water;
- Designing energy and lighting systems to conserve fossil fuels and maximize the use of renewable resources;
- Selecting materials that are nontoxic, biodegradable, and easily recycled and that minimize the impacts on landfills and otherwise reduce waste; and

[2]LEED for Schools is under development in 2006, in collaboration with CHPS.

- Creating an indoor environment that provides occupants with a comfortable temperature, and good air quality, lighting, and acoustics.

Green school guidelines also recommend construction techniques to meet objectives such as the appropriate storage of materials on construction sites to avoid water damage, the reduction of waste materials and appropriate disposal to reduce resource depletion, and the introduction of commissioning practices[3] to ensure the performance of building systems. Operation and maintenance practices to achieve good indoor environmental quality include using nontoxic cleaning products, replacing air filters in ventilation systems regularly, and establishing a long-term indoor environmental management plan.

Because they follow conventional design and construction practice, current green school guidelines typically treat materials, lighting, ventilation systems, windows, and other building components as individual elements, not as interrelated systems. They allow designers to accumulate credits by optimizing certain components or systems (e.g., energy efficiency) while suboptimizing or ignoring others (health and development). In doing so, such guidelines fail to account for the interrelationships among systems, occupants, and ongoing practices. Nor do they identify the potential need for compromises or trade-offs among design objectives in order to optimize overall building performance as it relates to multiple objectives.

In 2005, the Massachusetts Technology Collaborative (MASSTECH) promulgated draft guidelines for green school construction and major renovations in Massachusetts based on both the California CHPS and the most recent LEED® standards. MASSTECH defined a high-performance green school as having three distinct attributes:

- It is less costly to operate than a conventional school;
- It is designed to enhance the learning and working environment; and
- It conserves important resources such as energy and water.

[3]Commissioning is "a quality-focused process for enhancing the delivery of a project. The process focuses on verifying and documenting that the facility and all of its systems and assemblies are planned, designed, installed, tested, operated to meet the Owner's Project Requirements" (ANSI, 2004, p. 1). Commissioning is discussed in detail in Chapter 9.

STATEMENT OF TASK

At the request of MASSTECH, the Barr Foundation, the Kendall Foundation, the Connecticut Clean Energy Fund, and the USGBC, the National Research Council (NRC), through the Board on Infrastructure and the Constructed Environment (BICE), appointed the Committee to Review and Assess the Health and Productivity Benefits of Green Schools. The committee's charge was to "review, assess, and synthesize the results of available studies on green schools and determine the theoretical and methodological basis for the effects of green schools on student learning and teacher productivity." In the course of the study, the committee was asked to do the following:

1. Review and assess existing empirical and theoretical studies regarding the possible connections between the characteristics of "green schools" and the health and productivity of students and teachers.
2. Develop an evaluation framework for assessing the relevance and validity of individual reports that considers the possible influence of such factors as error, bias, confounding, or chance on the reported results and that integrates the overall evidence within and between diverse types of studies.
3. Report the results of this effort in a manner that will facilitate the identification of causal relationships and the subsequent implementation of beneficial practices.
4. Identify avenues of research that represent potentially valuable opportunities to leverage existing knowledge into a better understanding of the relationships between green building technologies in schools and the performance of students and teachers.

The committee, appointed in January 2005, was composed of experts in education, green building technology, student performance, sustainable design, indoor environments, buildings and health, epidemiology, materials, infectious diseases, school design, management, and administration, and research methodology. The committee held five 2- or 3-day meetings between April 2005 and January 2006 and was briefed by the study sponsors, by experts in building design and operation, and by researchers. The committee also toured the Capuano green school in Somerville, Massachusetts, and held several conference calls. An interim report was published in January 2006 (http://fermat.nap.edu/catalog/11574.html).

FINDING AND RECOMMENDATIONS

Finding 1: School buildings are composed of many interrelated systems. A school building's overall performance is a function of interactions among these systems, of interactions with building occupants, and of operations and maintenance practices. However, school buildings are typically designed by specifying individual components or subsystems, an approach that may not recognize such interactions. Current green school guidelines reflect this approach and, in so doing, they allow for buildings focused on specific objectives (e.g., energy efficiency) at the expense of overall building performance.

Recommendation 1a: Future green school guidelines should place greater emphasis on building systems, their interrelationships, and overall performance. Where possible, future guidelines should identify potential interactions between building systems, occupants, and operation and maintenance practices and identify conflicts that will necessitate trade-offs among building features to meet differing objectives.

Recommendation 1b: Future green school guidelines should place greater emphasis on operations and maintenance practices over the lifetime of a building. Systems that are durable, robust, and easily installed, operated, and maintained should be encouraged.

ORGANIZATION OF THE REPORT

Although the committee has already emphasized the need for a systems approach to buildings, the report is generally organized by various building features. This construct is necessary because research on the built environment, like building design, typically focuses on specific features and systems, as opposed to overall performance. Following the discussion in Chapter 2, "Complexity of the Task and the Committee's Approach," Chapters 3 through 7 discuss specific elements of indoor environmental quality and their effects on human health, learning, and productivity. Each chapter begins by identifying issues, then discusses design requirements and solutions related to those issues, and ends with a discussion of current green school guidelines. These discussions are generalized and primarily based on the committee's review of the CHPS, Washington State, and draft MASSTECH guidelines and supporting documents.

Chapter 3, "Building Envelope, Moisture Management, and Health," focuses on excess moisture or dampness in buildings, its effects on human health, and the building envelope as a source of excess moisture.

Chapter 4, "Indoor Air Quality, Health, and Performance," discusses indoor air quality as a function of pollutant loads, temperature, humidity, and ventilation rates, and how these factors are related to human health and development.

Chapter 5, "Lighting and Human Performance," focuses on lighting quality as it affects both the visual and circadian systems, the relationship of light to learning and development, and sources of light and their various qualities.

Chapter 6, "Acoustical Quality, Student Learning, and Teacher Health," describes how noise levels affect speech intelligibility and student learning, and how excessive noise may relate to voice impairment among teachers.

Chapter 7, "Building Characteristics and the Spread of Infectious Diseases," addresses the issue of common infectious diseases (colds and flu and other viral infections) and building interventions that can help to interrupt the transmission of these diseases.

Chapter 8, "Overall Building Condition and Student Achievement," summarizes the findings of a number of published and unpublished studies looking at overall conditions and functionality of school buildings and their effects on student achievement.

Chapter 9, "Processes and Practices for Planning and Maintaining Green Schools," highlights the importance of participatory planning, setting up commissioning processes, monitoring building performance, postoccupancy evaluations, and training for educators and support staff.

Chapter 10, "Linking Green Schools to Health and Productivity: Research Considerations," summarizes the committee's specific suggestions for future research on green schools and discusses factors to be considered in designing such research.

Because indoor environmental quality is a composite of air quality, light and noise levels, temperature, humidity, and other factors and a result of the interactions of various building systems, some topics are discussed in more than one chapter. Thus, excess moisture in buildings and its relation to health is discussed in Chapters 3 (building envelope), 4 (indoor air quality) and 7 (building characteristics and the spread of infectious diseases).

NOTE TO READERS

Research literature from a diverse set of disciplines and from U.S. and international studies is reviewed and cited in this report. Two systems of measurement, English and metric, were used in the original studies. For example, some research on ventilation rates uses liters per person per second as a measure, while other research uses cubic feet per minute or air exchanges per hour. The committee provides explanations of the measures as appropriate. However, it has not attempted to convert such measures to a common standard, because doing so could result in inadvertent errors that would detract from this report and its findings.

2

Complexity of the Task and the Committee's Approach

The study committee was charged to review and assess existing empirical and theoretical studies on the possible connections between characteristics of green schools and the health and performance of students and teachers. A number of factors make this task more complex than might first be evident. These include lack of a clear definition of what constitutes a green school; variations in current green school guidelines; the difficulty of measuring educational and productivity outcomes; the variability and quality of the research literature; and confounding factors that make it difficult to separate out the effects of building design, operations, and maintenance from the effects of other variables. Each of these factors is discussed below to provide context for the committee's approach to meeting its charge.

COMPLEXITY OF THE TASK

Lack of Clear Guidance About the Characteristics of Green Schools

Ideally, the committee's evaluation would be based on a single clear definition of a green school and/or guidance about the characteristics of green schools that differentiate them from conventional schools. However, as noted in Chapter 1, green school-related definitions and guidelines vary in the objectives to be achieved, individual elements, and level of detail. Some green school guidelines, such as California's, go beyond environmental objectives, specifying objectives for occupant health and productivity. This variation in definitions and guidelines is understandable given the grass roots nature of the green building movement and the fact that the guidelines were developed by a diversity of organizations. However, it did complicate the committee's task.

In reviewing examples of green school objectives and guidelines for California and Washington State and draft guidelines for Massachusetts, the committee determined that such documents have two complementary, but not identical, goals: (1) to support the health and development (physical, social, and intellectual) of students, teachers, and other staff by providing a healthy, safe, comfortable, and functional physical environment and (2) to have positive outcomes for the environment and the community. Thus, research on green schools might be conceptualized as having two quite different outcomes: improved student and staff health and development or improved environment and community. Because they were first developed to minimize adverse environmental impacts, current green school guidelines place less emphasis on supporting human health and development. Accordingly, and in line with its charge, the committee focused on outcomes associated with student and teacher health, learning, and productivity.

Measuring Educational and Productivity Outcomes

An assessment of the effects of a school building on student learning and health and teacher health and productivity must be set in context. First, time spent in school is, at most, 40 to 50 hours per week, and other environments, including home, neighborhood, recreational, cultural, and religious settings, could equally affect health and performance. In addition, differing populations and individuals may have differing sensitivities and responses to features or attributes of the built environment.

Education, which is the transfer of knowledge and skills to people (learning), is also difficult to measure directly. Learning is influenced by many factors, including the quality of curriculum, teacher education/experience, parental support, peer support, student background, quality of the administration, instructional materials, laboratory equipment, and educational standards. Policy researchers suggest that teaching and learning might be shaped by various state policies and their implementation regarding teacher education, licensing, hiring, and professional development and by national policies such as the No Child Left Behind Act (P.L. 107-110).

Measuring productivity is equally difficult. Productivity for an individual or an organization has been defined as the quantity and/or the quality of the product or service delivered (Boyce et al., 2003). Productivity is influenced by both the individual and the system within which he or she works. Increasing evidence is available to indicate that the built environment can influence both individual and organizational productivity, which can be measured, in some cases, by the number of units manufactured or the number of words typed correctly in a given amount of time. Absenteeism is often used as a surrogate for productivity, the rationale being that people who are absent are less productive than people who are on the job.

For work like teaching that cannot be measured in units, productivity is more difficult to define and measure. Productivity is closely linked to the "quality" of teachers, both individually and collectively. Researchers have used various measures such as educational preparation, certification/licensure, and scores on teacher examinations (Praxis is one) to try to objectively evaluate the quality of teaching staff. These measures can be quantified to a certain extent and are used quite efficiently in research studies. They have not been tested sufficiently, however, to enable a great deal of confidence in their validity. For instance, a teacher's years of experience is to a degree a measure of quality, because a teacher is usually evaluated in conjunction with a decision about the granting of tenure. To that extent, an administration may judge a teacher's quality to be high enough, given his or her satisfactory performance during the preceding years, that it decides to grant tenure. Conversely, there is no guarantee that someone gaining tenure is doing anything more than meeting the minimum standards. Also, the basis for awarding tenure is not consistent across school districts. Thus, the same level of performance might gain a teacher tenure in one school district but not in another.

Researchers cannot directly access an individual's personal experiences but must gain access to them indirectly by measuring outcomes such as learning, productivity, or attendance or asking them to reveal their preferences by means of surveys. In the studies reviewed by the committee, multiple and quite varied measures were used. One measure, absenteeism, was used as a surrogate measure for student health, teacher health, student learning, and teacher productivity (Table 2.1).

Because of the multiplicity of variables that can influence learning, research that evaluates educational interventions does not often find a large effect on student achievement outcomes. Interventions such as intensive in-service education for teachers, major revisions of curriculum, and addition of curricular units on substance abuse prevention or economics tend to show small gains in student achievement at best.

Similarly, it is very difficult to draw firm inferences about cause-and-effect relationships between a physical intervention—say, school design—and health and performance outcomes for individuals. Thus, it is not unexpected if research on building characteristics or operation practices finds only a small effect on health, learning, or productivity.

TABLE 2.1 Examples of Indicators Used to Measure Relevant Outcomes

Indicator	For Students	For Teachers
Health	Asthma Allergies Cold/flu and other respiratory diseases Headaches Absenteeism	Asthma Allergies Cold/flu and other respiratory diseases Vocal fatigue Headaches Absenteeism
Development (learning and productivity)	Standardized test scores Sustained attention Working memory Prospective memory Reading comprehension Verbal comprehension Demonstration of concepts Disciplinary incidents Absenteeism	Attitude/motivation Teaching behaviors/methods Knowledge Teacher examination scores (Praxis) Absenteeism

The Variability and Quality of Research

The committee's task was further complicated by gaps in the research literature. For example, much of what is known about the impact of the physical environment on health and performance is based on studies of adult populations. Some research is quite narrow because it makes an effort to control for all possible extraneous variables. Other research is so broad that confounding variables are possible, if not likely.

In general, much less is known about the impact of the school environment—green or conventional—on children's health and learning as compared with the impact of workplace environments on adults' health and productivity. Extrapolating from studies of adults to draw conclusions about a much younger population can be suspect because children are still developing physically, mentally, and emotionally. In addition, environmental factors in a school may interact with genetic factors to determine the degree to which a child develops language skills or displays asthma symptoms.

An additional challenge is the great variability in the characteristics and standards used by different disciplines to conduct research. For example, medical research may employ clinical trials and intervention studies in which various factors can be controlled for and their results directly evaluated using established protocols. Often clinical trials of drugs include placebos administered according to the same protocol as the drug of interest. Epidemiological studies often rely on statistical significance as a quantitative measure of the extent to which chance or sampling variation might be responsible for an observed association between an exposure and an adverse event. In these studies, quantitative estimation is firmly founded in statistical theory on the basis of repeated sampling.

Social science research testing links between the built environment and the behaviors of occupants, in contrast, cannot set up strictly controlled trials or easily manipulate variables to test for statistical significance. Studies attempting to link health, students' learning, and teachers' productivity with school environments cannot control for the effects of other, nonschool environments. In addition, there are no standard protocols for conducting building-related research, although some studies have used similar methodologies or evaluation methods, including multiple regression analyses and measures of statistical significance.

Confounding Factors

The committee was asked to consider the possible influence of confounding, bias, error, and chance in the relevant literature. Confounding, in particular, poses a major challenge to researchers and those evaluating their work. Confounding occurs when the evaluation of a relationship between two features or groups is biased by a third factor. For example, in any epidemiological study comparing an exposed group with a nonexposed group, a simple comparison of the groups may exaggerate the true difference or hide it, because it is likely that the two groups will differ with respect to factors that are also associated with the risk of the outcome of interest, such as socioeconomic status. Said differently, a simple comparison of the incidence of health outcomes among exposed and nonexposed groups may exaggerate an apparent difference, because socioeconomic status is also thought to influence the incidence of several health problems.

A variety of confounding factors will be present in any study that attempts to link features of school buildings with student and teacher health and development. Age differences among students will influence the outcomes of research into health and learning: Young children inhale more air per pound of body weight than teenagers; their tissues and organs are actively growing; they are still developing language and cognitive skills; and they spend more time in school than anywhere else but home. Extrapolating findings from studies of adults and applying those findings to children and teenagers can also be problematic.

Research studies that include measures of student learning in making comparisons to other variables must try to control extraneous variables. For instance, when comparing student achievement test scores with the condition of the physical environment, a researcher would typically attempt to control other variables that influence outcomes, including the socioeconomic status of the students or teachers and the curriculum. Student test scores will also be influenced by conditions in the classroom on the day of the test and other factors, like the amount of time spent on the task.

Another confounding factor is that school buildings themselves are not standardized, making direct comparisons between school environments problematic. Unlike the houses in tract developments, most schools are designed as unique structures, whose features depend on the resources available, construction methods, curriculums, populations, and building codes in effect at the time. They may have one or multiple stories, may accommodate a hundred or several thousand students, may be 5 or 100 years old; may be designed with open plans or traditional classrooms; may use central air handling systems or natural ventilation.

The condition of the school buildings is also a factor: School conditions diverge widely from one part of the country to another or even within localities. Those school systems with fewer financial resources and a greater percentage of students from low-income households seem to have school buildings in worse condition than those school systems that have a high tax base and the financial resources to solve the problems of providing good school buildings. In addition, it is difficult to find data about the condition of school buildings: The Educational Writers' Association (1989) found that few states were able to properly evaluate school buildings because departments of education had few or no personnel to conduct such evaluations. Consequently, data collection was erratic.

Another confounding factor is that building systems and characteristics operate in an integrated fashion to effectively deliver (or not) overall building performance, whose components include thermal comfort; air, visual, acoustic, and spatial quality; and long-term building integrity. Building performance will be affected by the operation and maintenance of these integrated systems over time and by the occupants of buildings and their activities.

Interactions between people and built environments are numerous and difficult to account for in research studies. For instance, the physical and psychological health of teachers may be affected by a school building's characteristics and conditions. When teachers' well-being suffers, so too might their instruction and their participation in school activities. In that sense, teachers can be seen as both an outcome of building conditions (they might be healthier or not, more motivated or not) and a mediating variable (they might teach differently or interact with students differently depending on their health and well-being) in explaining students' experiences. They may also positively or negatively alter a school's physical setting by adjusting the temperature of a room, opening or closing windows, and other actions.

Similarly, student learning and health might be directly or indirectly related to a school building's condition: If there is high absenteeism at a school among teachers because of poor air quality, the quality of instruction may suffer. If students are absent because of poor air quality, they have less opportunity to receive instruction. And, just as teachers can alter the school environment, so too can students.

A third population that can affect and be affected by a school's environment is the administrators (principals, financial staff, counselors, librarians) and support staff (building operations and maintenance personnel, cleaning crews, and kitchen workers). These groups may spend as much time in a school building as teachers and students and sometimes at different times of the day (before and after classes, on weekends, during school breaks and summer vacation). The quality of the support staff training may significantly affect the performance of building systems, the timeliness and quality of maintenance and repair, and cleaning practices.

With outcomes as complex as student health and learning—influenced as they are by many individual, family, and community factors—it might be theoretically possible to design research that controls for all potentially confounding variables, but it would be difficult to conduct such research.

THE COMMITTEE'S APPROACH

The line of reasoning inherent in this study's task—mapping connections from physical environments to student and teacher outcomes—poses significant challenges, as described above. An additional challenge is the directionality of relationships. Green schools might have positive effects on student health, but it might also be that students who live in communities inclined to build green schools are healthier to begin with. Another challenge is that the effects of physical environments might be trumped in some way by other forces—teacher quality, parental involvement, or financial resources, for instance.

To help inform the committee about mechanisms by which the physical environment might affect student learning, teacher productivity, or the health of students and teachers, the committee developed a conceptual model for evaluating links between green school buildings and outcomes for learning, health, and productivity (Figure 2.1).

In the conceptual model, one underlying assumption is that a green school's location and design (site, orientation, envelope, heating, ventilation, air conditioning, acoustics, lighting) will result in an indoor environment with appropriate (or inappropriate) levels of moisture, ventilation,

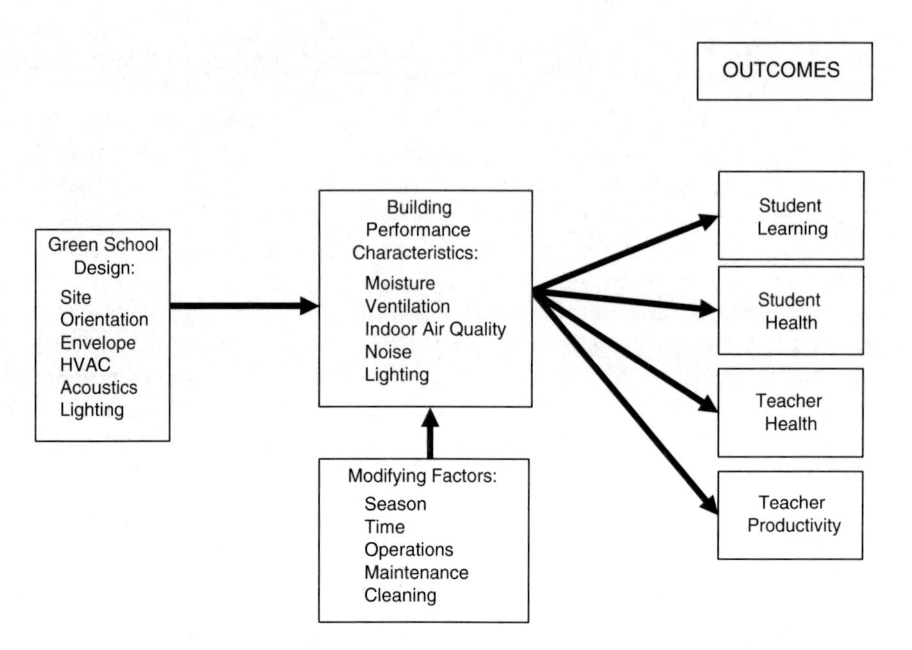

FIGURE 2.1 Conceptual model for evaluating links between green school buildings and outcomes for learning, health, and productivity.

air quality, noise, light, and other qualities. It also assumes that the indoor environment will be modified by season (e.g., presence of airborne pollen) and over time (e.g., mold growth from chronic water leakage) and by operational, maintenance, repair, and cleaning practices. Finally, the model assumes that the indoor environment can affect student learning and health and teacher health and productivity.

The committee's initial review of the literature focused specifically on identifying studies that purport to address the connections between green school design, student learning and health, and teacher productivity and health. No well-designed, evidence-based studies concerning the *overall* effects of green schools on the health or development of students and teachers were identified. The committee also did not identify any evidence-based studies that analyze whether green schools differ from conventional schools in regard to occupant health and productivity. A few studies were identified that examined specific building features often emphasized in green school design and the effects of these features on health and performance. Among those are a series of studies on daylighting and student achievement produced between 1999 and 2003. These studies are discussed in Chapter 5. For the most part, however, the literature on green schools, health, and productivity consists of anecdotal information and case studies of varying quality.

This lack of well-designed, evidence-based studies specifically related to green schools is understandable, because the concept of green schools is relatively new and evidence-based studies require a significant commitment of time and resources. Furthermore, other research on the factors that influence student learning does not typically look at factors related to building design and maintenance.

A much more robust body of scientific evidence is available that looks at building characteristics emphasized in green school design—envelope, mechanical and engineering systems, lighting, acoustics—and their relationships to occupant health, development, and productivity. Typically, such studies look at a single system or a very limited number of variables, and their quality varies. For example, a review of the literature on building characteristics, dampness, and health effects identified 590 epidemiological studies addressing these topics. Of these, only 61 met standards for strong study design and the provision of useful information (NORDAMP, 2002). Relatively few studies look at two or three building systems and their effects. With these caveats in mind, the committee determined it should review the scientific literature relating to those characteristics that are typically emphasized in definitions of and guidelines for green schools, such as indoor air quality, and building characteristics for which significant scientific research was available.

However, a review of all research literature that touches on some aspect of buildings and their potential impacts on occupant health, learning, or productivity is an undertaking beyond the resources of this study. Where rigorous reviews of a particular aspect of interest have been conducted by the National Research Council or other organizations or researchers, the committee relied on that work (e.g., IOM, 2004; Mendell and Heath, 2004). Where research is fairly limited but important to the study, the committee conducted its own review. Ultimately, the scope of the literature review was based on the committee's collective judgment as to where its efforts should be concentrated to best address the task statement and meet the sponsors' requirements. In all cases, the committee describes the source of the literature reviewed, the research methodology used, and the basis for the committee's conclusions.

The committee's examination of the evidence included assessing the relevance and validity of individual studies, integrating the overall evidence within and between diverse types of studies and across studies of different building characteristics, and then formulating its findings and recommendations. Because those tasks required thoughtful consideration of the evidence and could not be accomplished by adherence to a narrowly prescribed formula, the committee's approach evolved throughout the study process and was determined to some degree by the nature of the available evidence. Ultimately, the committee used its collective best judgment with regard to evaluating the published research and the plausibility of explanations of physiological mechanisms, and then integrated the results of those studies found to be useful (even if flawed) into its findings, conclusions, and recommendations.

In its deliberations, the committee determined that given the complexity of interactions between people and their environments, it may never be possible to categorically establish a causal relationship between an attribute of a school building and its effect on students, teachers, and staff. The effects of the built environment will necessarily appear to be small, given the large number of variables. Nor may it be possible to quantify the effects of one feature, such as acoustics, on student learning. However, the committee believes that empirical measures do not necessarily capture all relevant considerations that should be applied when evaluating research results. Qualitative aspects of the environment are also important. Thus, in the committee's collective judgment, there is value in attempting to identify design features and building processes and practices for green schools that may lead to improvements in learning, health, and productivity for students, teachers, and support staff, even if the empirical results are less than robust.

FINDINGS AND RECOMMENDATION

Finding 2a: Given the complexity of interactions between people and their environments, establishing cause-and-effect relationships between an attribute of a green school or other building and its effect on people is very difficult. The effects of the built environment may appear to be small given the large number of variables and confounding factors involved.

Finding 2b: The committee did not identify any well-designed, evidence-based studies concerning the *overall* effects of green schools on human health, learning, or productivity or any evidence-based studies that analyze whether green schools are actually different from conventional schools in regard to these outcomes. This is understandable because the concept of green schools is relatively new, and evidence-based studies require a significant commitment of resources.

Finding 2c: Scientific research related to the effects of green schools on children and adults will be difficult to conduct until the physical characteristics that differentiate green from conventional schools are clearly specified. With outcomes as complex as student and teacher health, student learning, and teacher productivity influenced by many individual family and community factors, it may be possible in theory to design research that controls for all potentially confounding factors, but difficult in practice to conduct such research.

Recommendation 2: The attributes of a green school that may potentially affect student and teacher health, student learning, and teacher productivity differently than those in conventional schools should be clearly specified. Once specified, it may be possible to design appropriate research studies to determine whether and how these attributes affect human health, learning, and productivity.

3

Building Envelope, Moisture Management, and Health

Excess moisture or dampness and mold growth in buildings have been associated with some upper respiratory symptoms (nasal congestion, sneezing, runny or itchy nose) and respiratory diseases, especially asthma, in children and adults (IOM, 2000, 2004). Asthma affects 8 to 10 percent of the population and even larger proportions of children in certain cities or in poor urban populations. It is a common cause of absence from school and from the workplace as well; 14 million days of school loss were recorded in 1994-1996, 3.4 days per child with asthma (Cox-Ganser et al., 2005).

As long as a building is properly designed, sited, constructed, operated, and maintained, excess moisture can be managed effectively. However, excess water or moisture in a building can lead to structural failures and health problems when materials stay wet long enough for microbial growth, physical deterioration, or chemical reactions to occur (IOM, 2004).

Unfortunately, moisture problems in buildings are common in all climates of the United States. It is generally accepted that more than 75 percent of all building envelope (foundation, walls, windows, roof) problems are caused by excess moisture. Moisture also significantly affects the comfort and health of occupants (Lstiburek and Carmody, 1994; Achenbach, 1994; Tye, 1994; Deal et al., 1998).

Moisture in buildings comes from both outdoor and indoor sources: rain, snowmelt, groundwater, construction materials, plumbing systems, kitchens, shower rooms, swimming pools, and wet surfaces such as mopped floors (Figure 3.1). People also bring in rain and snow on their

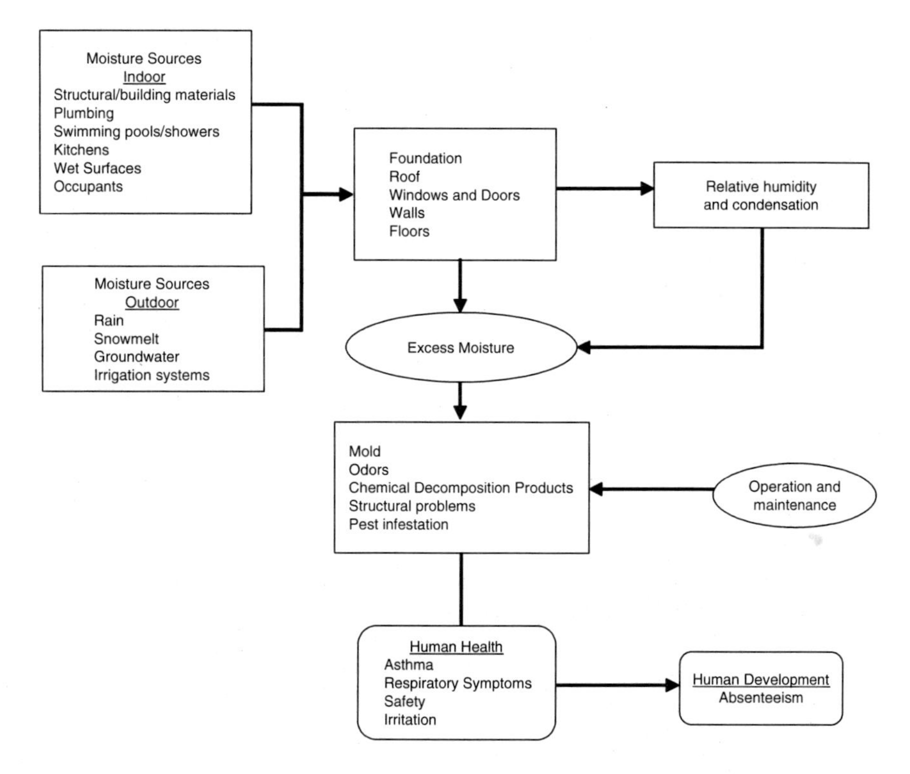

FIGURE 3.1 Conceptual model of excess moisture and its potential impacts on students, teachers, and support staff.

clothing. The effects of excess moisture are manifested by mold, mildew, rotted wood, insect infestations, spalling of masonry, condensation on surfaces, stained finishes, peeling paint, and reduced service life of materials and systems.[1]

Typically, current green school guidelines address moisture management issues as they relate to the siting of a building (they discourage development on wetlands or below the 100-year floodplain), the placement of drainage and irrigation systems to prevent water accumulation in or near buildings, keeping construction materials dry, and using walk-off mats and grills to prevent the buildup of snow and rain brought into a building by people. The committee believes that moisture control over a school

[1]Excess moisture is also related to indoor air quality and heating, ventilation, and air-conditioning systems, and surfaces. These subjects are addressed in Chapters 4 and 7.

building's life cycle is critical and should have a more prominent place in future green school guidelines, for reasons outlined in this chapter.

EXCESS MOISTURE AND HEALTH

Recently concerns have been raised that indoor moisture, dampness, and mold growth can lead to a variety of health problems in adults and children. Scientific research indicates that the most consistent and convincing associations relate to respiratory disease, especially asthma.

Asthma is a disorder in which the airflow is obstructed. People with asthma are subject to episodic wheezing, coughing, and shortness of breath. Although these symptoms are common clinical features of asthma, they are common symptoms of other respiratory illnesses as well. Finding a widely accepted definition of this disease has proved problematic, but the following has been offered by the Institute of Medicine as the most acceptable:

> Asthma is understood to be a chronic disease of the airways characterized by an inflammatory response involving many cell types. Both genetic and environmental factors appear to play important roles in the initiation and continuation of the inflammation. Although the inflammatory response may vary from one patient to another, the symptoms are often episodic and usually include wheezing, breathlessness, chest tightness, and coughing. Symptoms may occur at any time of the day, but are more commonly seen at night. These symptoms are associated with widespread airflow obstruction that is at least partially reversible with pharmacologic agent or time. Many persons with asthma also have varying degrees of bronchial hyperresponsiveness. Research has shown that after long periods of time this inflammation may cause a gradual alteration or remodeling of the architecture of the lungs that cannot be reversed with therapy (IOM, 2000, pp. 23-24).

Indoor environments are an important factor in chronic asthma symptoms and morbidity (the incidence of disease), whether these environments are in the home, workplace, or school. The Institute of Medicine (IOM) has issued two reports on the association between excess moisture or dampness and mold growth on the one hand and respiratory illness in building occupants on the other: *Clearing the Air: Asthma and Indoor Air Exposures* (IOM, 2000) and *Damp Indoor Spaces and Health* (IOM, 2004). Both concluded that damp, moldy buildings were associated with respiratory symptoms both in people suffering from chronic asthma and in the general population. To the extent that green schools can be designed to minimize the contribution of dampness to the incidence of asthma, they can have a positive impact on human health.

The report *Damp Indoor Spaces and Health* considered separately the common respiratory symptoms (wheeze, cough, shortness of breath) and the diagnosis of asthma (usually based on reported physician diagnosis, reversible obstruction measured by lung function tests, or the respondent's use of appropriate medication). In addition, the report distinguished asthma development (the appearance of asthma for the first time) from asthma exacerbations (asthma symptoms in persons with chronic asthma). The report found sufficient evidence of an association between indoor dampness and several respiratory health outcomes, including asthma. That is, an association has been observed between indoor dampness and respiratory health outcomes in studies in which chance, bias, and confounding can be ruled out with reasonable confidence. However, the evidence was not strong enough to say there was a causal relationship, i.e., that dampness directly caused the respiratory health outcomes.

There are at least two distinct variants of asthma: an extrinsic, allergic variant that occurs in the context of immunoglobulin E (IgE)-mediated sensitization to environmental allergens and an intrinsic, nonallergic variant with no detectable sensitization and low IgE concentrations. In both variants, the airways are strikingly hyperresponsive, and symptoms may also be mediated by irritant responses. In those studies that evaluated asthmatic patients for IgE-mediated sensitization, the association was stronger in sensitized individuals; thus the IOM study concluded that the association was strongest in sensitized individuals. In addition, studies of the general population consistently found an association between dampness and mold and the symptoms of cough or wheeze. Because asthma has been diagnosed in only 8 to 10 percent of the population, it was unlikely that this relationship could be accounted for in these studies by asthma alone. Thus, the 2004 IOM study concluded that moisture and mold were also associated with cough and wheeze in the general population.

Not enough studies were found to support an association between dampness and mold and the *development* of asthma. Nine of the 10 studies available found an association with moisture, mold, or both. Only 1 (Jaakola et al., 2002) found that the association was insignificant. Particularly important were three birth cohort studies (Belanger et al., 2003; Slezak et al., 1998; Maier et al., 1997) in which infants and children who were genetically at risk of developing asthma were observed for several years. Stark et al. (2003) reported on a birth cohort of 849 infants less than 1 year old who had at least one sister or brother with physician-diagnosed asthma; they found that wheeze and persistent cough were associated with airborne concentrations of *Penicillium* and *Cladosporium*, two types of mold commonly found in indoor air samples.

Finally, upper respiratory symptoms (nasal congestion, sneezing, runny or itchy nose) were also associated with damp indoor environments and mold. Like asthma, chronic rhinitis (inflammation of the inner lining of the nose) has allergic and nonallergic variants. The allergic variant occurs in the context of IgE-mediated sensitization to environmental allergens. In the studies included in the 2004 IOM report, upper respiratory symptoms were associated with dampness and mold in persons with self-identified allergic rhinitis as well as in the general population. Other studies reported that the frequency of "colds," that is, of acute viral infectious rhinitis, was associated with dampness and mold. Because the cause of upper respiratory symptoms could not be identified, the committee concluded that the symptoms, but not a specific illness, were associated with dampness and mold.

The mechanisms by which damp indoor spaces and mold are associated with respiratory illness are not clear, but there are several possibilities. First, many people with asthma demonstrate IgE-mediated sensitization to mold, so the symptoms could be related to specific immune mechanisms. Mold produces a number of materials, such as peptidoglycans and polysaccharides, that induce inflammation through the innate immune pathways. Other materials such as volatile organic compounds[2] and toxins may have direct effects because asthmatic airways are excessively responsive to exposures to irritants. Other organisms such as gram-negative or gram-positive bacteria might coexist with mold in damp environments; endotoxin or lipoteichoic acid from these organisms might induce airway symptoms. The interaction of moisture and mold with building materials may produce metabolites that have direct irritant effects on asthmatic airways.

The findings of key relevance to this report from *Damp Indoor Spaces and Health* are summarized in Box 3.1.

Health effects from excess moisture are mediated by increased indoor organisms or by deteriorated building materials that produce bioaerosols, defined as contaminants that come from living organisms and are airborne. Health effects may arise from a wide variety of mechanisms, including direct irritation of the eye or respiratory mucosa, immunologic bacteria, direct inflammation induced by toxic effects of bacterial products such as endotoxin or by VOCs, immunologic sensitization and inflammation, and the direct effects of fungal exotoxins. The multiplicity of possible mechanisms illustrates that the pathways to respiratory effects will be complex, but all of these mechanisms are plausible consequences of excess indoor moisture. Maintaining structures that are dry (i.e., without excess moisture) could prevent all of these effects.

[2]Volatile organic compounds (VOCs) are emitted as gases from certain solids or liquids. VOCs include a variety of chemicals, some of which may have short- and long-term adverse health effects.

BOX 3.1
Findings from the Institute of Medicine Relevant to This Study

Sufficient Evidence of an Association Between Dampness and Respiratory Health: Evidence is sufficient to conclude that there is an association. That is, an association between the agent and the outcome has been observed in studies in which chance, bias, and confounding can be ruled out with reasonable confidence.

- Asthma symptoms in sensitized persons
- Cough
- Wheeze
- Upper respiratory (nasal and throat) tract symptoms

Limited or Suggestive Evidence of an Association: Evidence is suggestive of an association between the agent and the outcome but is limited because chance, bias, and confounding could not be ruled out with confidence. For example, at least one high-quality study shows a positive association, but the results of other studies are inconsistent.

- Asthma development
- Dyspnea (shortness of breath)
- Lower respiratory illness in otherwise healthy children

Sufficient Evidence of an Association Between Mold or Other Agents and Respiratory Health

- Asthma symptoms in sensitized persons
- Cough
- Wheeze
- Upper respiratory (nose and throat) tract symptoms
- Hypersensitivity pneumonitis in susceptible individuals

Limited or Suggestive Evidence of an Association

- Lower respiratory illness in otherwise healthy children

SOURCE: IOM (2004).

BUILDING ENVELOPE AND MOISTURE MANAGEMENT

The foundation, walls, windows, and roof of a building make up an "envelope" intended to shelter people, equipment, and furnishings from the weather and from natural and manmade hazards. Windows and doors allow outside air, light, people, equipment, and supplies to enter or exit a building. Skylights allow in natural light or daylight. Building envelopes can be designed for natural ventilation, for mechanically conditioned air systems, or for some combination of these. Whether planned or not, buildings have multiple openings that allow the penetration and internal movement of air, water, and contaminants.

Building assemblies exist in a dynamic environment. Some materials have the ability to store moisture and subsequently dry without harmful affects. Excess moisture can be controlled when a healthy balance or equilibrium is maintained between the rates of entry and removal. For example, masonry construction incorporating a drain screen in the walls can provide effective moisture control through a balance of storage capacity and high drying potential. Problems with building assemblies arise only when assemblies accumulate moisture faster than their ability to store and/or dry without associated degradation of performance: Because steel framing and gypsum wallboard have virtually no storage capacity, a small leak can quickly become a large problem.

A complex set of moisture-transport processes related to climate, building design, construction, operation, and maintenance determine whether a building will have excess moisture that could influence the health of the occupants. The approach used by many building scientists to understand and diagnose moisture transport is termed "source–path–driving force" analysis: For any particular case, there is a source of moisture, a pathway moisture follows, and a force that drives moisture along that pathway. If a building designer is able to control at least one of the three elements in this chain, moisture can be effectively controlled. Controlling more than one element provides for a valuable redundancy.

Effective moisture management considers the potential damage and degree of risk associated with each of the following four transport mechanisms (from most to least potent):

1. Bulk transport,
2. Capillary transport,
3. Air transport, and
4. Vapor diffusion.

Bulk transport is the liquid flow of rain, snowmelt, or groundwater into a building envelope under the influence of pressure differences exerted by gravity, hydrostatic pressure, wind, or air pressure. It is the most significant moisture transport mechanism that must be addressed by designers.

Capillarity is the wicking of liquid through the pore structure of a material (Lstiburek and Carmody, 1994; Straube, 2002). Wood, concrete, brick, and mortar are able to draw water into their porous structures in a manner similar to a sponge. Below-grade building assemblies like foundation walls, footings, and slabs are particularly sensitive because they are in contact with wet soil and standing water. Water drawn through these assemblies evaporates into the inside space, elevating interior humidity levels. Above-grade components are at capillarity risk too. Rain and splashback on exterior walls can be drawn into the envelope through capillary pathways, such as overlaps in siding, pores in wood and masonry materials, and joints between otherwise nonporous materials.

Air transport is the transfer of water vapor through the movement of air. It poses a threat roughly equal to that posed by capillarity. Air moves moisture into and through building assemblies both from within the conditioned space and from the outside (Lstiburek and Carmody, 1994; Straube, 2002; Rousseau, 2003). Water is simply one of the many gases found in air. As air moves, so does the water vapor. Airborne moisture moves under the influence of air pressure differentials created by wind, mechanical equipment, and the stack effect.[3] The relationship between air transport of moisture and heating, ventilation, and air-conditioning (HVAC) systems is discussed in Chapter 4.

Vapor diffusion is the least powerful moisture transport mechanism (Lstiburek and Carmody, 1994; Straube, 2002; Achenbach, 1994) and is often confused with air transport because it too deals with invisible water vapor. The primary difference is that diffusion is the movement of water vapor *through* the actual structural matrix of a material, not through holes and cracks in an assembly, as is the case with air movement. Moreover, diffusion is driven by vapor pressure, not air pressure. As a result the process is slow and relatively weak when compared with air transport of water vapor. The directional vector is typically from the warmest side of the building envelope toward a colder side. The rate of diffusion is a function of the vapor permeability of a material and the driving force, vapor pressure.

[3]One way to ventilate a building that is hotter or colder on the inside than outside is to use what is known as "stack effect." Because of the temperature difference, the air inside the building is either more or less dense than the air outside. If there is an opening high in the building and another low in the building, a natural flow will be caused. If the air in the building is warmer than the air outside, this warmer air will float out the top opening, being replaced with cooler air from outside. If the air inside is cooler than that outside, the cooler air will drain out the low opening, being replaced with warmer air from outside.

Newly constructed buildings give off significant amounts of moisture during the first 2 years of use. Materials like concrete, masonry, lumber, plaster, and various surface coatings hold large quantities of water that evaporate into the indoor air.[4] These "wet" assemblies can be designed to dry to the outdoors to reduce the loading of internal air. Alternatively, the moisture contained in the indoor air can be diluted through ventilation or dehumidification (Lstiburek and Carmody, 1994).

SOLUTIONS/DESIGN REQUIREMENTS FOR MOISTURE MANAGEMENT

Designing for moisture management is complicated. Architects must incorporate design features to control the entry of large amounts of rain and ground water from the outside, block capillary transport through and within the structure, and prevent excessive transport of water vapor through air movement and vapor diffusion. Compounding the challenge, designers must concurrently provide a healthy indoor environment, minimize energy use, and control construction costs (Lstiburek and Carmody, 1994; Achenbach, 1994; Powell, 1994). In addition, moisture control must be placed in the context of structural design, operation, maintenance, and use of the building as it relates to external conditions including climate, soil, and microclimates (IOM, 2004; Powell, 1994; Lstiburek and Carmody, 1994).

A central role for the above-grade building envelope is to keep rainwater out. If rainwater is allowed to penetrate the envelope, it will overwhelm the impact of all other interior moisture sources—for example, moisture from occupants, plants, bathrooms, kitchens, and evaporation from wet surfaces (Christian, 1994). Architects can provide for redundant layers of protection, include a drainage plane within the envelope, and attempt to control air pressure differentials across the exterior cladding with a vented rain screen (Lstiburek and Carmody, 1994).

Water originating above-grade from rain, surface runoff, or snowmelt is typically drawn down through the soil by gravity. Migrating surface water and subsurface water from a high water table can enter below-grade building envelopes through the cracks, joints, holes, or pores in a structure. Subsurface transport is driven by hydrostatic pressure, gravity, and/or capillary transport (Lstiburek and Carmody, 1994; Christian, 1994; Straube, 2002). Proper site selection, the use of appropriate drainage systems, and application of effective barrier membranes are the most successful bulk-moisture control strategies for below-grade assemblies.

[4]For example: poured concrete releases 146 pints per cubic yard during the 2-year period following construction (Christian, 1994).

In general, the easiest way to control capillary transport is to reduce the availability of moisture, seal the pores, make the pores too large to support capillarity and/or provide a receptor (Lstiburek and Carmody, 1994). Capillary transfer of below-grade moisture into concrete and masonry walls and floors can be controlled by the use of damp proofing, waterproofing membranes, and non-capillary-conducting drainage materials (Lstiburek and Carmody, 1994; ORNL, 1988; Tye, 1994).

One large source of moisture in building enclosures is the migration of moisture from the surrounding soil into foundations and basements. This moisture ultimately moves by air transport when wet surfaces evaporate into the air of conditioned spaces (Christian, 1994). Blocking this liquid source from entry before it becomes an airborne problem is a smart design choice. Installation of air-barrier systems at the interior and/or exterior surfaces of a building shell can help to reduce the amount of air and airborne moisture transported.

Moist air also leaks into basements through floor and slab perimeters, wall joints, cracks, windows, and around drains. Winter stack forces can act to depressurize the basement area, drawing moist air into the basement through these pathways. Moist air is subsequently directed to the upper regions of a structure. Since below-grade moisture is one of the most powerful sources of moisture in conditioned spaces, building dry foundation assemblies can be one of the most effective strategies used to control interior airborne moisture (Lstiburek and Carmody, 1994; ORNL, 1988; Christian, 1994).

The easiest way to control diffusion is by installing vapor impermeable materials on the side of the assembly with the highest vapor pressure (Rousseau, 2003). Building codes consider materials with a permeability rating of ≤1 to be vapor barriers. As a general rule, designers should position vapor barriers toward the inside surface in heating-dominated climates and toward the outside surface in cooling-dominated climates. Many building codes and architectural standards require seams and holes in vapor barriers to be sealed to form a continuous, uninterrupted line of protection. However, effectiveness depends on a material's vapor permeability and surface area covered. In other words, if 95 percent of an envelope surface is covered with a vapor barrier, the barrier is 95 percent effective as a vapor diffusion retarder (Lstiburek and Carmody, 1994). This relationship provides designers and builders with some leeway, suggesting as it does that diffusion barriers need not be installed perfectly. However, in order for air barriers to be effective, they must be continuous and durable.

In some cases diffusion can drive moisture through an envelope from the inside and from the outside. A structure built in a locality where a balanced mix of heating and cooling is required is one example. A foundation assembly where significant amounts of moisture exist inside and outside the envelope is another example of a situation where moisture can diffuse in both directions. And porous claddings like masonry and wood that become rain soaked and subsequently exposed to solar radiation give rise to vapor driving inward, even in a heating-dominated climate (Lstiburek and Carmody, 1994). This type of diffusion will affect the design of vapor barriers. Buildings should be designed such that moisture can dry toward both the inside and the outside.

Box 3.2 lists design measures that could be incorporated into green school guidelines to ensure appropriate moisture management as it relates to a building's envelope. Excellent resources for proper moisture control design include *The Moisture Control Handbook: Principles and Practices for Residential and Small Commercial Buildings* by Joseph Lstiburek and John

BOX 3.2
Design Strategies for Moisture Control

- Site building thoughtfully, considering exposures imposed by sun, wind, standing water, runoff, and seasonal high water.
- Avoid building on sites below the 100-year floodplain or in wetlands.
- Direct water away from foundations using gutters, downspouts, overhangs, and site grading. Slope soil away from building at a 5 percent grade.
- Use free-draining backfill or manufactured drainage mats adjacent to below-grade surfaces. Connect free-draining backfill or mats to a subgrade drainpipe system that directs water through a filter to block fines (particles smaller than the average) from entering the drainage system.
- Install a layer of impervious soil or cap material around the perimeter of the building to prevent surface water from entering the below-grade drainage system.
- Install moisture proofing on all below-grade surfaces to serve as a capillary break between the soil and below-grade assembly. Provide a capillary break between the footing and foundation wall.
- Install a vapor-impermeable membrane below the slab to interrupt the flow of moisture from the soil to the slab.
- Install a ≥4-inch gravel drainage layer below membrane-protected slabs and footings.
- Control rain penetration into above-grade walls using rain screens, drainage planes, building papers, and appropriate placement of flashings.

Carmody (1994), *The Building Foundation Design Handbook* (ORNL, 1988), and *Moisture Control in Buildings* (ASTM, 1994). Building commissioning can also be an effective way to identify and preempt potential moisture problems in schools. Building commissioning is discussed in greater detail in Chapter 9.

In addition to bringing potential health benefits, designing for effective moisture management will probably have benefits for the building itself. The more durable a building is, the longer its components will last (Deal et al., 1998). Materials in long-lived building assemblies are replaced less frequently than those in nondurable structures. This makes dry structures resource- and energy-efficient, because no replacement materials need be harvested, mined, or produced, nor is energy used to make, transport, or assemble the replacement components. Dry buildings also require fewer

- Where irrigation systems are being used, install them such that water is not directly splashed onto building walls or foundations.
- Install drainage systems to prevent the accumulation of water under or near buildings.
- Include an air barrier on the inside and/or outside of building assemblies as appropriate.
- Include a vapor barrier for roof, wall, and foundation assemblies as appropriate to the local climate.
- Incorporate water-shedding systems, waterproof membranes, and drainage systems in roofing assemblies according to guidelines published annually by the National Roofing Contractors Association (NRCA), Rosemont, Illinois.
- Ventilate roof assemblies in all but airtight, heavily insulated assemblies in the harshest climates.
- Construct building envelopes that can store water if they get wet and are able to dry if they get wet, as appropriate for the local climate.
- Install thermal insulation in a position to control dew point locations in the building envelope when possible and as appropriate for climate and building assembly.
- Minimize the amount of moisture contained in the original building materials when possible.
- Keep construction materials dry and well-ventilated during the construction process; discard any materials showing signs of mold or mildew.
- Use walk-off grills and mats to prevent the buildup of moisture and rain brought inside the building by students, teachers, support staff, or visitors.

resources and money for repair and maintenance. For example, damp surfaces cause stains and peeling paint, which necessitate frequent repainting and cleaning. For these reasons, dry buildings may have lower life-cycle costs, in addition to offering potential health benefits for occupants.

More research is needed on the moisture resistance and durability of materials used in school construction. Such research should also investigate other properties of these materials, such as their generation of bioaerosols and indoor pollutants as well as the environmental impacts of producing and disposing of them.

CURRENT GREEN SCHOOL GUIDELINES

Green school guidelines and standards endeavor to prevent rain and snow from entering a building in a number of ways. These include discouraging development on sites below the 100-year flood plain or in wetlands; keeping site irrigation to a minimum; designing all drainage systems and HVAC condensate drainage systems to prevent the accumulation of water under, in, or near buildings; using walk-off grills and mats to prevent the buildup of moisture from rain and snow brought in by occupants; keeping all construction materials dry and well ventilated; discarding materials that have been wet for more than 24 hours; and requiring a maintenance plan that specifies the staff time and materials that will be dedicated to HVAC, plumbing, and roof systems.

FINDINGS AND RECOMMENDATIONS

Finding 3a: There is sufficient scientific evidence to establish an association between excess moisture, dampness, and mold in buildings and adverse health outcomes, particularly asthma and respiratory symptoms, among children and adults.

Finding 3b: Excess moisture in buildings can lead to structural damage, degraded performance of building systems and components, and cosmetic damage, all of which may result in increased maintenance and repair costs.

Finding 3c: Well-designed, -constructed, and -maintained building envelopes are critical to the control and prevention of excess moisture and molds. Designing for effective moisture management may also have benefits for the building, such as lower life-cycle costs.

Finding 3d: Current green school guidelines typically do not adequately address the design detailing, construction, and long-term maintenance of buildings to ensure that excess moisture is controlled and a building is kept dry during its service life.

Recommendation 3a: Future green school guidelines should emphasize the control of excess moisture, dampness, and mold to protect the health of children and adults in schools and to protect a building's structural integrity. Such guidelines should specifically address moisture control as it relates to the design, construction, operation, and maintenance of a school building's envelope (foundations, walls, windows, and roofs) and related items such as siting and landscaping.

Recommendation 3b: Research should be conducted on the moisture resistance and durability of materials used in school construction. Such research should also investigate other properties of these materials such as the generation of bioaerosols and indoor pollutants as well as the environmental impacts of producing and disposing of these materials.

4

Indoor Air Quality, Health, and Performance

Indoor air quality, which is a function of outdoor and indoor air pollutants, thermal comfort, and sensory loads (odors, "freshness"), can affect the health of children and adults and may affect student learning and teacher productivity.

Pollutants are generated from many sources. Outdoor pollutants include ozone, which has been associated with absenteeism among students. Pollutants and allergens in indoor air—mold, dust, pet dander, bacterial and fungal products, volatile organic compounds, and particulate matter—are associated with asthma and other respiratory symptoms and with a set of building-related symptoms (eye, nose, and throat irritations; headaches; fatigue; difficulty breathing; itching; and dry, irritated skin). In some cases, outdoor pollutants react with indoor chemicals to create new irritants.

Thermal comfort is influenced by temperature, relative humidity, and perceived air quality (sensory loads) and has been linked to student achievement as measured by task performance. Relative humidity is also a factor in the survival rates of viruses, bacteria, and fungi and their effects on human health (see Chapter 7, "Building Characteristics and the Spread of Infectious Diseases").

Heating, ventilation, and air-conditioning (HVAC) systems are intended to provide (1) effective outside air delivery to rapidly dilute or filter out air contaminants and (2) thermal comfort for building occupants by heating or cooling outside air coming into occupied spaces. Ventilation can be supplied through mechanical systems, which draw air into and push air out of a building, or "naturally," through the opening and closing of doors and windows and by uncontrolled leakage points through a building's envelope. A variety of mechanical systems is available, including hybrid systems that use both natural and mechanical ventilation.

HVAC systems must be properly designed and sized to handle the sensible and latent heat loads of outside and recirculated air. If not properly designed, operated, and maintained, HVAC systems can themselves generate pollutants and excess moisture, thereby affecting the health of occupants. The principal standards and guidelines for HVAC system design and operation in the United States are (1) American Society of Heating, Refrigeration, and Air-Conditioning Engineers (ASHRAE) Standard 62.1-2004, "Ventilation for Acceptable Indoor Air Quality"; (2) American National Standards Institute (ANSI)/ASHRAE Standard 55-2004, "Thermal Environmental Conditions for Human Occupancy"; (3) the Department of Energy's EnergySmart Schools guidelines; and (4) individual state codes, some of which are based on or refer to the International Building Code or other codes. Because industry standards for ventilation and energy efficiency have been developed separately, they have, in some cases, had the net effect of increasing relative indoor humidity.

As shown in Figure 4.1, the complex interactions between indoor and outdoor pollutants, moisture/humidity, HVAC systems, operations and maintenance practices can affect occupants' health, comfort, and productivity. These topics are discussed in greater detail in the rest of the chapter.

POLLUTANT SOURCES

Pollutants are generated by many sources both internal and external to a school. External sources include combustion products; biological material; and particulate matter and ozone entering through air intakes and the building envelope. People themselves can carry pollen and allergen sources, such as dust mites and pet dander, into a school on their shoes, skin, and clothes. Internal sources include but are not limited to combustion products; building materials and equipment; educational materials; cleaning products; biological agents; and human activity. In some cases, outdoor pollutants react with indoor chemicals to produce new irritants.

Outdoor Sources of Pollutants

Outdoor air pollutants can affect the health of children and adults in two ways. First, students, teachers, administrators, and support staff are exposed to outdoor pollutants before they enter a building, which can lead to increased respiratory symptoms (Schwartz, 2004). Second, outdoor

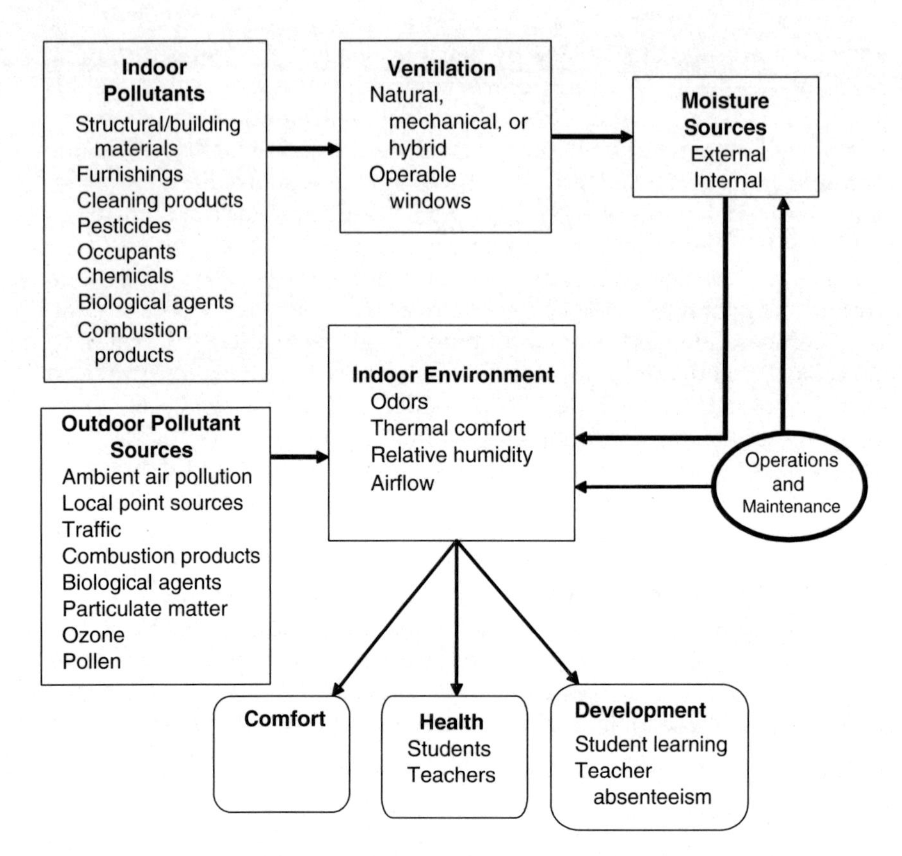

FIGURE 4.1 Relationships between pollutants, moisture, and ventilation and human comfort, health, and development.

sources of pollution can contribute to indoor air pollutant concentrations when outdoor air is drawn into a school building through air intakes located at the rooftop, at ground level, or from below-grade "wells." Outside air also enters the building through doors, windows, ventilation shafts, and leaks in the building envelope.

Mendell and Heath (2004, p. 9) found that "a substantial literature of strongly designed cohort studies is available on associations between outdoor pollutants and attendance of children at school." They concluded that there was strongly suggestive evidence that absence from school increased with exposure to ozone at higher concentrations. However, the findings were mixed on the associations of school absence with exposure to outdoor nitrogen oxides, carbon dioxide, and particles <10 μm.

Site location can be an important determinant of outdoor pollutants. Schools next to high-traffic areas or with school buses idling their engines next to school doorways, windows, and air intakes may have higher levels of outdoor air pollutants being drawn indoors (Park and Jo, 2004; Blondeau et al., 2005; Singer et al., 2004; Behrentz et al., 2005).

Other significant sources of outdoor pollutants are plant-derived materials, or biomass, which can generate bioaerosols, including molds, fungi, and pollen. An IOM study (2002, p. 8) found as follows:

> Although there is sufficient evidence to conclude that pollen exposure is associated with exacerbation of existing asthma in sensitized individuals, and pollen allergens have been documented in both dust and indoor air, there is inadequate or insufficient information to determine whether indoor air exposure to pollen is associated with exacerbation of asthma.

The IOM study also noted that "there is relatively little information on the impact of ventilation and air cleaning measures on indoor pollen levels, although it is clear that shutting windows and other measures that limit the entry rate of unfiltered air can be effective" (p. 14).

Indoor Sources of Pollutants

Indoor pollutants include chemicals, allergens, volatile organic compounds (VOCs), particulate matter, and biological particles or organisms. Chemicals in indoor environments include combustion products such as nitrogen oxides (NO_x), sulfur oxides (SO_x), and carbon monoxide (CO). Combustion products can be generated by gas-fired pilot lights in kitchens and laboratories. Other sources of indoor chemical pollutants include building materials (e.g., structural materials such as particleboard, adhesives, insulation); furnishings (carpets, paints, furniture); products used in a building (cleaning materials, pesticides, markers, art supplies); and equipment (copiers and printers).

Indoor allergen sources—house dust mites, pet dander, cockroaches, rodents, and seasonal pollens—can be brought into a building by occupants, can be generated by furry animals kept in classrooms, or can be attracted to food sources in, for example, school kitchens and cafeterias. Daisey et al. (2003) found that a variety of bioaerosols (primarily molds and fungi, dust mites, and animal antigens) could be found in school environments.

Volatile organic compounds (VOCs) and semivolatile organic compounds (SVOCs) are chemical compounds used extensively in building materials such as adhesives for wood products and structural materials, paints, and carpet adhesives. They also are found in art supplies, paints and lacquers, paint strippers, cleaning supplies, pesticides, office equipment such as copiers and printers, correction fluids and carbonless copy paper, graphics and craft materials, markers, and photographic solutions. In fact, there are no places in schools where VOCs and SVOCs are not found.

Outdoor sources of VOCs and SVOCs include fuels and combustion, biological organisms, and pesticides. Research has shown that concentrations of VOCs are consistently higher indoors than outdoors (Adgate et al., 2004; Wallace, 1991), and studies in homes suggest that indoor concentrations vary depending on the specific VOC (Weisel et al., 2005; Meng et al., 2005). One study also showed that building renovation contributes significantly to total VOC concentrations (Crump et al., 2005).

Particulate matter (PM) includes solid particles ranging in size from ultrafine (<0.1 μm) to relatively large (>10 μm). These particles come from outdoors (including dusts and particles from traffic, stationary sources, and microorganisms), and indoors (humans, building materials, fibers, bioaerosols, mold, pet dander) (Afshari et al., 2005). Larger PM remains suspended in air for relatively short periods of time, instead settling on floors, surfaces, and furnishings. Smaller PM has longer suspension times—i.e., it remains airborne longer. Particulate matter has been implicated in a number of health effects, primarily respiratory and cardiac (Nel, 2005). Particulate matter can absorb VOCs, which may affect occupants' health and comfort (Nilsson et al., 2004).

Larger PM tends to be related to housekeeping practices, ineffective filtration by HVAC systems, and local activity. Finer PM tends to be more independent of these factors, and a fraction of finer PM will even diffuse through structures and so be not removable by HVAC filtration.

One important group of PM is the airborne allergens, including molds and fungi, dander and other body fragments, dust mites, and cockroach antigens. Because these bioaerosols can induce an immune response, they are capable of causing illness at very low exposure levels and also of causing more severe respiratory disease than PM from nonbiological sources. The strength of the association of each of these bioaerosols with illness was summarized in *Clearing the Air: Asthma and Indoor Air Exposures* (IOM, 2000), and many of them were found to be more strongly related to asthmatic symptoms than were moisture and mold.

Improperly maintained HVAC systems can themselves be a source of pollutants. Several findings in *Damp Indoor Spaces and Health* (IOM, 2004) pertain specifically to the design and operation of HVAC systems as a critical factor in the control of moisture and mold growth in buildings:

- Although relatively little attention has been directed to dampness and mold growth in HVAC systems, there is evidence of associated health effects (p. 42).

- Liquid water is often present at several locations in or near commercial-building HVAC systems, facilitating the growth of microorganisms that may contribute to symptoms or illnesses (p. 42).
- Microbial contamination of HVAC systems has been reported in many case studies and investigated in a few multibuilding efforts (p. 43).
- Sites of reported contamination include outside air louvers, mixing boxes (where outside air mixes with recirculated air), filters, cooling coils, cooling coil drain pans, humidifiers, and duct surfaces (p. 43).
- Bioaerosols from contaminated sites in an HVAC system may be transported to occupants and deposited on previously clean surfaces, making microbial contamination of HVAC systems a potential risk factor for adverse health effects (p. 43).

The Menzies et al. study (2003) of ultraviolet germicidal irradiation (UVGI) of drip pans and cooling coils indicates that limiting the microbial contamination of HVAC systems may yield health benefits. The use of UGVI is discussed in greater detail in Chapter 7.

Indoor Air Chemistry

Ozone (O_3) is a primary pulmonary irritant that also plays an important role in indoor chemistry. Although ozone concentrations are generally higher outdoors than indoors, indoor ozone concentrations can be appreciable, infiltrating a building through windows, doors, and the envelope (Weschler et al., 1992). Ozone concentrations might be expected to be higher in naturally ventilated buildings. Indoor ozone sources include printers, copiers, and electrostatic air cleaners if they are not adequately maintained or are improperly exhausted. Sources of indoor terpenes and other unsaturated hydrocarbons are numerous and include cleaning products and air fresheners (Nazaroff and Weschler, 2004).

Reactions among reactive gases (such as ozone) and commonly occurring, nonirritating organic compounds (certain terpenes such as limonene and pinene) can generate products that are highly irritating and can impact human health and comfort (Karlberg et al., 1992; Weschler and Shields, 1997). The process of these ongoing reactions has been termed "indoor air chemistry" (Weschler et al., 1992). Ozone/terpene reaction products have been shown to cause greater airway irritation than either original product (Wolkoff et al., 2006; Weschler, 2004).

Using chamber studies, Weschler et al. (1992) demonstrated the formation of these reaction products from carpets. Weschler later discovered that concentrations of products generated by reactions among indoor pollutants increased as ventilation decreased (Weschler and Shields, 2000). This increase in reaction products is independent of the diurnal variation in ozone levels or of outside ozone levels. These results suggest that maintaining adequate ventilation rates may reduce the potential for reactions among airborne pollutants that generate even more reactive and irritating products.

VENTILATION

Ventilation rate is based on the outdoor air requirements of a ventilation system. Ventilation effectiveness is based on the ability of the system to distribute conditioned air within occupied spaces to dilute and remove air contaminants. The principal standard for ventilation rates is American Society of Heating, Refrigeration, and Air Conditioning Engineers (ASHRAE) Standard 62.1-2004, "Ventilation for Acceptable Indoor Air Quality." However, Daisey et al. (2003), in a comprehensive review of the literature related to indoor air quality, ventilation, and health symptoms in schools, found that reported ventilation and carbon dioxide (CO_2) levels indicated that a significant proportion of classrooms did not meet (then) ASHRAE Standard 62-1999 for minimum ventilation rate.[1]

A number of studies in schools have reviewed the effect of ventilation rates on health, productivity, and airborne pollutant control. Typically, these studies also look at a second variable, such as temperature or humidity, both of which are components of thermal comfort, to identify any confounding or synergistic effects (Figure 4.2).

School-Related Studies

Wargocki et al. (2005) conducted a field intervention experiment in two classes of 10-year-old children. Average air temperatures were reduced from 23.6°C to 20°C, and outdoor air supply rates were increased from 5.2 to 9.6 liters per second (L/s) per person in a 2 × 2 crossover design, each condition lasting a week. Tasks representing eight different aspects of schoolwork, from reading to mathematics, were performed during appropriate lessons, and the children marked visual-analogue scales each week to indicate their perception of building-related symptom intensity. In this study, increased ventilation rates corresponded to increased task completion in multiplication, addition, number checking ($p < .05$), and subtraction ($p < .06$). Reduced temperature corresponded to increased task completion in subtraction and reading ($p < .001$) and fewer errors in check-

[1]The ASHRAE standard has since been revised to require even higher ventilation rates.

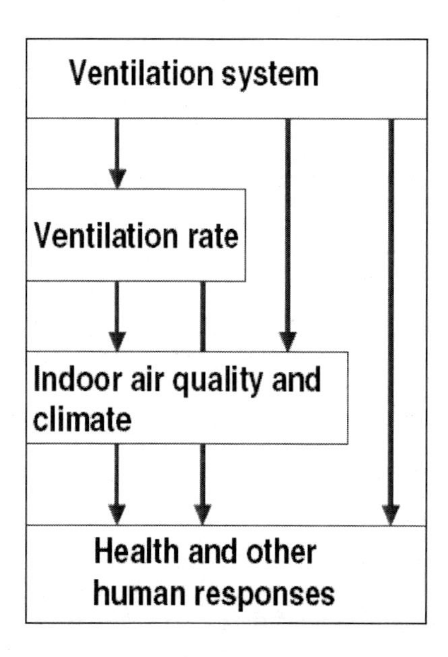

FIGURE 4.2 Pathways by which ventilation affects health and other human responses. SOURCE: Sepännen and Fisk (2004a).

ing a transcript against a recorded voice reading aloud ($p < .07$). When reduced temperature was combined with increased ventilation rates, task completion increased in a test of logical thinking ($p < .03$). Experimental data indicated that increasing ventilation rates from 5.4 to 9.6 L/s per person and decreasing temperatures from 24°C to 20°C could improve the performance of schoolwork by children as measured by task completion.

Smedje and Norbäck (2000) investigated the impact of improving school ventilation systems on allergies, asthma, and asthma symptoms in schoolchildren. They issued questionnaires to 1,476 children in 39 schools (mixed primary and secondary schools) from 1993 to 1995. Various exposure factors were measured in 100 classrooms during this time. In 12 percent of the classrooms, new ventilation systems were installed; their effect was to increase the air-exchange rate and reduce humidity. Air pollutant levels were lower in classrooms with the new ventilation systems. This investigation indicated a health improvement for children in the classrooms with increased ventilation, lower humidity, and reduced airborne pollutants. The incidence of asthma symptoms, but not allergies, was reduced in the classrooms with the new ventilation systems.

Shendell et al. (2004b) explored the association between student absences and indoor CO_2 levels. These researchers noted that since measuring the actual ventilation rate is expensive and potentially problematic, the indoor concentration of CO_2 has often been used as a surrogate for the ventilation rate per occupant, including in schools. They measured the short-term (5 min) CO_2 levels in 409 traditional and 25 portable classrooms from 22 schools in Washington State. Attendance data were collected from school records. Their results indicated that a 1,000 parts per million (ppm) increase above the outdoor concentration of CO_2 was associated with statistically significant 10 to 20 percent increases in student absences.

Mendell and Heath (2004) looked at the literature and found

- Suggestive, although not fully consistent, evidence linking low outdoor ventilation rates in buildings to decreased performance in children and adults and
- Suggestive but inconsistent evidence linking lower ventilation rates with decreased attendance among adults.

No studies on the impact of increased ventilation rates and/or effectiveness on teacher health and productivity were identified. However, multiple studies have looked at the effects of increased ventilation rates in office buildings and call centers on adult health and productivity (Milton et al., 2000; Seppänen et al., 1999; Seppänen and Fisk, 2004b, 2005; Fisk et al., 2003; Wargocki et al., 2002; Wyon, 2004). These studies are discussed below.

Studies of Offices and Other Building Types

In a study of 3,720 hourly employees of a large Massachusetts manufacturer in 40 buildings with 115 independently ventilated working areas, Milton et al. (2000, p. 212) analyzed the relationship between the rate at which outdoor air was supplied for ventilation and the amount of sick leave taken. The researchers found "consistent associations of increased sick leave with lower levels of outdoor air supply and IEQ [indoor environmental quality] complaints." Seppänen and Fisk (2004a) developed a further quantitative relationship by fitting the data from these epidemiological studies using the Wells-Riley model of airborne disease transmission to predict the relationship. The model predicted that there would be a decrease in illness over time with increased ventilation rates.

Seppänen et al. (1999) reviewed the literature on the association of ventilation rates in nonresidential and nonindustrial buildings (primarily offices) with health and performance outcomes. The review included 20 studies investigating the association of ventilation rates with human responses and 21 studies investigating the association of CO_2 levels with human responses. A majority of studies found that ventilation rates of less than 10 L/s per person were associated in all building types with a statistically significant worsening in one or more health or perceived air quality outcomes. Some studies found that increasing ventilation rates up to 20 L/s per person was associated with significant decreases in the prevalence of building-related symptoms or with further significant improvements in perceived air quality. The ventilation rate studies reported relative risks of 1.1 for building-related symptoms at low ventilation rates and 6 at high ventilation rates.

The report *Clearing the Air* (IOM, 2000, pp. 15-16) stated:

> There are both theoretical evidence and limited empirical data indicating that feasible modifications in ventilation rates can decrease or increase concentrations of some of the indoor pollutants associated with asthma by up to approximately 75%. Limited or suggestive evidence exists to indicate that particle air cleaning is associated with a reduction in the exacerbation of asthma symptoms. . . . It should also be noted that microorganisms can grow on some air-cleaning equipment such as filter media; thus improperly maintained air cleaners are a source of indoor pollutants.

The same report also stated:

> Control options for chemical and particulate pollutants in indoor environments include source modification (removal, substitution, or emission reduction), ventilation (exhaust or dilution), or pollutant removal (filtration). The various forms of pollutant source modification are usually the most effective. For most gaseous pollutants—NO_2 for example—removal via air cleaning is not presently practical (p. 15).

Ventilation and health in nonindustrial indoor environments were the subjects of a European Multidisciplinary Scientific Consensus Meeting (EUROVEN) review of the scientific literature on the effects of ventilation on health, comfort, and productivity in offices, schools, homes, and other nonindustrial environments (Wargocki et al., 2002). The group reviewed 105 papers and judged 30 as being conclusive, providing sufficient evidence on ventilation, health effects, data processing, and reporting. The EUROVEN group agreed that ventilation is strongly associated with comfort (perceived air quality) and health (building-related symptoms, inflammation, infections, asthma, allergy) and found an association between ventilation and productivity (performance of office work). It concluded that increasing outdoor air supply rates in nonindustrial environments improved the perceived air quality. The EUROVEN group also concluded that outdoor air supply rates of less than 25 L/s per person increased the risk of building-related symptoms, increased the use of short-term sick leave, and led to decreased productivity among office building occupants.

Wyon (2004) investigated the productivity of office workers when indoor air pollutant loads were reduced by removing pollutant sources or increasing ventilation rate. He showed that during realistic experimental exposures lasting up to 5 hours, the performance of simulated office work was increased 6 to 9 percent by the removal of common indoor sources of air pollution, such as floor-coverings and old air filters, or by keeping the sources in place while increasing the clean air ventilation rate from 3 to 30 L/s per person. Wyon then went on to confirm these laboratory findings in a field investigation during an 8-week period. He concluded that reduction in pollutant loads in buildings can be expected to reduce building-related symptoms.

A recent study of relocatable classrooms found that with either conventional or alternative building materials, ventilation could reduce VOC concentrations to less than 1 ppb (Hodgson et al., 2004). In a study by Reitzig et al. (1998) VOCs were measured in the air of 51 renovated rooms in different types of buildings—private apartments, schools, kindergartens, and office buildings. The only common characteristic was that all of the rooms had been renovated within the last 2 years and complaints had been received about the quality of the indoor air. The investigation found that modern ecological building materials contained less volatile and less common substances but with increased indoor persistence that could partially account for the increasing number of complaints in relation to building-related symptoms.

Seppänen and Fisk (2005) reviewed the scientific literature regarding the effects of ventilation on indoor air quality and health, focusing on office-type buildings. Overall their literature review indicated that ventilation has a significant impact on several important user outcomes, including:

- Communicable respiratory illnesses,[2]
- Building-related symptoms,
- Task performance and productivity,
- Perceived air quality among occupants and sensory panels, and
- Respiratory allergies and asthma.

Overall, these studies strongly indicate that although compliance with ASHRAE standards for ventilation rates may be the minimum acceptable

[2]See Chapter 7.

standard, increasing the ventilation rate beyond the ASHRAE standard will further improve indoor air quality, comfort, and productivity and may have health benefits (Wargocki et al., 2002, 2005; Smedje and Nörback, 2000; Shendell et al., 2004a,b).[3] The incremental effect of an increased ventilation rate may be attributable to a reduction in the pollutant load to which building occupants are exposed. However, the research conducted to date has not established an upper limit on the ventilation rates, above which the benefits of outside air begin to decline.

THERMAL COMFORT

Human perception of the thermal environment depends on four parameters: air temperature, radiant temperature, relative humidity, and air speed (Kwok, 2000). Perception is modified by personal metabolic rates and the insulation value of clothing. Thermal comfort standards are essentially based on a set of air and radiant temperatures and relative humidity levels that will satisfy at least 80 percent of the occupants at specified metabolic rates and clothing values.

There is a robust literature on the effects of temperature and humidity on occupant comfort and productivity, primarily from studies in office buildings (Fanger, 2000; Sepännen and Fisk, 2005; Wyon, 2004; Wang et al., 2005). These studies show that productivity declines if temperatures go too high (Federspiel et al., 2004). However, there is a paucity of studies investigating the relationship between room temperatures in schools and occupant comfort or productivity (Mendell and Heath, 2004).

ASHRAE has codified the air temperature, relative humidity, radiant temperature, and air movement conditions under which occupants should feel "thermally neutral." Guidance is found in ASHRAE Standard 55-2004, "Thermal Environmental Conditions for Human Occupancy," which provides a range of temperatures and relative humidity for winter and summer conditions. When applying current standards, several points are relevant to the school environment:

1. ASHRAE Standard 55-2004 and the ISO 7730 Standard for "Moderate Thermal Environments" are based on experimental studies of adults, not children.
2. New "adaptive" models of thermal comfort have not been incorporated into current standards used for the design of mechanical ventilation systems for schools. The metabolic rates of students

[3]On March 28, 2006, ASHRAE announced it will study the impact of ventilation rates on occupants' health. The project, Scientific Review of Existing Information Related to the Impact of Ventilation Related to Health, 1443-RP, is expected to take 18 months to complete.

vary across a school day as they engage in recess or lunch and move between rooms.

3. HVAC system design focuses almost exclusively on the thermal and humidity specifications as directed by building codes. Distribution of air diffusers is assumed to satisfy other requirements for air movement and prevention of thermal stratifications. Radiant heat gains and losses at a scale relevant to actual classroom utilization are not considered. Internal mixing, air velocities, and vertical temperature gradients are rarely explicit design considerations and are rarely assessed.

In addition, schools often have a higher occupancy density (more people per square foot of space) than office buildings. For these reasons and others, there is no assurance that school thermal conditions that meet current industry standards are optimal for student comfort or performance.

PERCEPTION OF AIR QUALITY (SENSORY LOADS)

An expanded definition of comfort includes the perception of air "quality." Occupants may perceive indoor air as heavy, stale, smelly, unpleasant, refreshing, or crisp. As the air is sensed, many attributes are integrated—its temperature, moisture content, odor, and chemical properties. Materials and educational supplies emit odorous compounds as do dirty filters and ducts, cleaning agents, kitchens, bathrooms, gymnasiums, art rooms, moldy surfaces, computers, and copying machines. Chemical reactions that occur indoors also give rise to particles and a host of odorous and irritating compounds (Weschler and Wells, 2005).

Fanger (2000) discusses perceived air quality and ventilation requirements in the context of indoor sensory pollution loads from occupants and materials. Exhaled breath, skin, sweat, dirty clothing, perfume, deodorants, and other body odors make the occupants themselves a source of the sensory pollution load degrading perceived indoor air quality. Using nonsmoking adults at 1 Met (metabolic rate) as a reference, kindergarten children at 2.7 Mets contribute 20 percent more to the sensory pollution load. Teenagers 14-16 years old at 1-2 Met activity levels contribute 30 percent more to the sensory pollution load that ventilation air has to handle to achieve the equivalent acceptance.

Wargocki et al. (1999) demonstrated that the use of low-emissions materials resulted in improved perceived air quality and productivity for typical office tasks and fewer reports of building-related symptoms. These findings were independently verified in the studies of Lagercrantz et al. (2000). Bakó-Biró et al. (2004) demonstrated that sensory pollution loads from common indoor objects like carpets, building materials, and personal computers decreased text typing performance as the percent of subjects dissatisfied with air quality increased. They reported a 0.8 percent decrease in text typing for a 10 percent decrease in perceived air quality. Wargocki et al. (2000) showed that increasing ventilation from 3 to 10 to 30 L/s per person improved simulated office work (typing rate and computation rate). These and other studies show that sensory pollution loads indoors are perceived by occupants and that dissatisfaction with perceived indoor air quality may have subtle effects on performance.

Moisture and relative humidity also play a role in the perception of air quality. Moisture in the air can lead to oxidation and chemical reactions by hydrolysis and decomposition, including enzymatic digestion by molds. These processes yield compounds that contribute to sensory pollution loads indoors. Fang et al. (1999a,b) found that perception depended on the enthalpy (heat content) of the air. Air that was cool and dry was perceived as "fresh" and "more pleasant" than air that was warm and moist. Figure 4.3, from Fang et al. (1999a,b) shows that in the absence of odorous sources people prefer air that is cooler and drier than the air commonly found indoors.

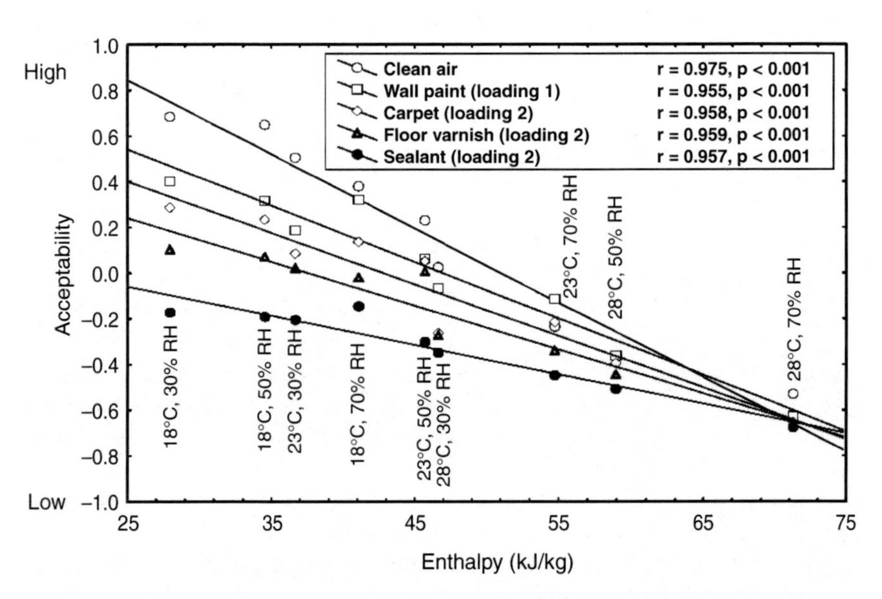

FIGURE 4.3 Acceptability of air quality as a function of enthalpy (heat content) and odorous sources. SOURCE: Adapted from Fang et al. (1999a,b).

The introduction of odor sources was perceived to degrade air quality whether introduced individually or in various combinations. Interestingly, when the enthalpy was high, objectionable odors could not be recognized as easily as when the enthalpy was low. In other words, people prefer cooler, drier air but are then more likely to detect odors, which diminish the perceived air quality.

Natural Versus Mechanical Ventilation

The available literature comparing natural to mechanical ventilation is inconclusive as it relates to health, with different authors reporting opposite results (Blondeau et al., 2005; Ribéron et al., 2002). Some evidence indicates that buildings with air conditioning systems may present an increased risk of building-related symptoms among occupants compared to buildings without air conditioning, and that improper maintenance, design, and functioning of air-conditioning systems contributes to increased prevalence of building-related symptoms among occupants (Figure 4.4) (Sepännen and Fisk, 2002; Wargocki et al., 2002; Graudenz et al., 2005).

A large European study that included a wide variety of building types, found that among the potential causes of adverse health effects due to HVAC systems were poor maintenance and hygiene in the ventilation systems, intermittent operation of the HVAC systems, lack of moisture control, lack of control of HVAC system materials, and dirty, loaded filters (Wargocki et al., 2002). This evidence would indicate that it is not so much the presence of HVAC systems that is a factor as it is the quality of HVAC system maintenance and operations.

Sepännen and Fisk (2002) reviewed the available office buildings literature about the relationship of ventilation system types and building-related symptoms and outlined the suspected risk factors of HVAC types and building features (Table 4.1).

Fan-assisted natural ventilation has been investigated in schools (Nördquist and Jensen, 2005) in climates without high humidities, although not in statistically controlled studies. In schools with openable windows, even the best-designed mechanical system may be operated as a hybrid system since teachers and staff often open windows and doors during pleasant weather.

Although there is good evidence that HVAC system characteristics can and do affect occupant health and comfort, including in schools, until recently there have been few studies that attempted to measure the magnitude of health or productivity effects. Sepännen and Fisk (2005) used previous studies (primarily in office buildings) to develop a model relating building ventilation rates, perceived air quality, and temperature

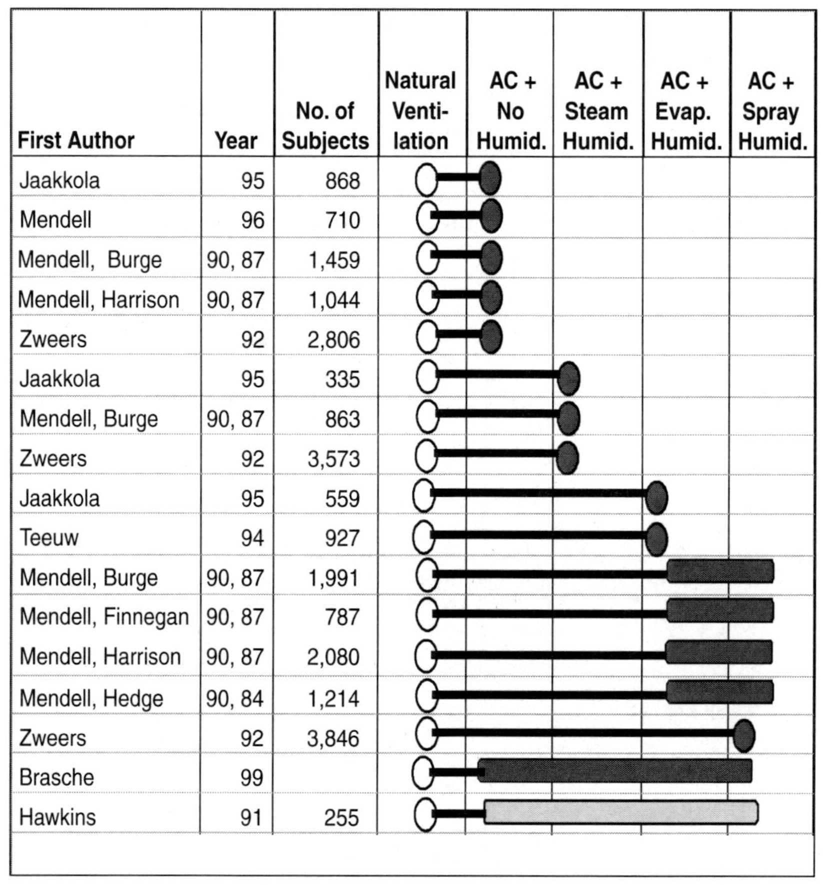

FIGURE 4.4 Studies comparing the health effects of buildings with and without air conditioning systems. SOURCE: Sepännen and Fisk (2002).

to occupant symptoms and productivity. They estimated that increasing the average ventilation rate from 0.45 to 1.0 exchange per hour[4] would reduce the sick leave used by office workers from 5 days per year to 3.9 days per year.

[4]Air exchange per hour is a ratio between the quantity of air delivered to a space in an hour divided by the volume of that space. It does not measure ventilation effectiveness for the removal of air pollutants.

TABLE 4.1 HVAC Systems and Risk Factors for Building-Related Symptoms

HVAC System Type	Risk
Natural ventilation with operable windows	No particle removal via filtration; poor indoor temperature and control; noise from outdoors; inability to control the pressure difference across the building envelope and exclude pollutant infiltration or penetration of moisture into structure; low ventilation rates in some weather; possible low ventilation rates in some portions of the occupied space.
Systems with ducts and fans but no cooling or humidification (simple mechanical ventilation)	HVAC components may be dirty when installed or become dirty and release pollutants and odors; poor control of indoor temperature due to absence of cooling; low humidity in winter in cold climates; high humidity during humid weather; noise generated by forced air flow and fans; draft caused by forced air flows.
Systems with ducts, fans, and cooling coils (air conditioning systems)	Additional risk factors from cooling coils: very high relative humidity or condensed moisture (e.g., in cooling coils and drain pans) and potential microbial growth; biocides used to treat wet surfaces such as drain pans and sometimes applied to nearby insulation.
Systems with ducts, fans, cooling coils, and humidifiers of various types	Additional risk from humidifiers: microbial growth in humidifiers; transport of water droplets downstream of humidifiers, causing wetting of surfaces; leakage and overflow of humidifier water; condensation from humid air; biocides in humidifiers; chemical water treatments in steam generators.
Systems with recirculation of return air (recirculation may occur in all mechanical HVAC systems)	Additional risks[a] from recirculation: indoor-generated pollutants are spread throughout the section of building served by the air-handling system; typically higher indoor air velocities increase risk of draft and HVAC noise; supply ducts and filters of HVAC system may become contaminated by recirculated indoor-generated pollutants.
Sealed or openable windows (windows may be sealed or openable with all types of mechanical HVAC systems)	Additional risk with sealed windows: no control of the environment if HVAC systems fails; psychological effect of isolation from outdoors. Additional risk with operable windows: more exposure to outdoor noise and pollutants.
Decentralized systems (cooling and heating coils located throughout building, rather than just in mechanical rooms)	Additional risk of decentralization: potentially poorer maintenance because components are more numerous or less accessible; potentially more equipment failures due to larger number of components.

[a]However, recirculation facilitates removal of indoor-generated pollutants using air cleaners, e.g., particle filters and may also decrease concentrations of pollutants near pollutant sources.

SOURCE: Sepännen and Fisk (2002).

VENTILATION SYSTEM STANDARDS

ASHRAE Standard committees periodically update the various standards documents. The ASHRAE Standard 62 series addresses ventilation in buildings and the 90.1 series addresses energy efficiency. After initially lowering ventilation requirements in response to the early 1970s energy crisis, ASHRAE Standard 62-2001, "Ventilation for Acceptable Indoor Air Quality," now requires substantially higher ventilation rates for schools and other buildings. ASHRAE Standard 90.1-2001, "Energy Standard for Buildings Except Low-Rise Residential Buildings," and other energy-saving measures such as more efficient motors, office equipment, and lighting, along with better thermal insulation for building envelopes, have systematically reduced the sensible heat loads of buildings. This has implications for HVAC design and operations.

In a report commissioned for AirXchange Corporation, TIAX (2003) demonstrated the consequence of systematic changes in buildings as a result of shedding heat loads and increasing ventilation. The net effect of lowering sensible heat loads while increasing ventilation rates without specifically dealing with latent heat loads has been to increase indoor relative humidity. Heat gains from within buildings have decreased. Henderson (2003) and Shirey (2003) report that in certain common conditions the cycling time of HVAC systems is shortened, leaving condensed moisture on coils that can reevaporate, adding moisture to the building supply air. Maintaining optimally comfortable humidity (between 40 and 50 percent) is more difficult. Higher humidity increases condensation on cooled indoor surfaces and thermal bridges. Humidity that remains above 65 percent for appreciable time increases the opportunity for mold growth.

Although there are studies looking at the energy efficiency and health effects of HVAC system operation, few if any studies directly compare the energy efficiency and health trade-offs, if any, of HVAC system operation (Engvall et al., 2005).

Calculations have shown that the estimated costs of poor indoor environmental quality is higher than the energy costs to heat and ventilate the same building (Seppänen, 1999). Additionally, studies show that many measures to improve indoor air quality are cost-effective when the health and productivity benefits of an improved indoor environment are included in the calculations (Fisk, 2000; Fisk et al., 2003; Seppänen and Vuolle, 2000; Tuomainen et al., 2003; Wargocki, 2003), although none of the studies was set in a school environment.

SOLUTIONS/DESIGN REQUIREMENTS FOR INDOOR AIR QUALITY

Ventilation

Ventilation systems are designed to manage the sensible and latent heat loads of buildings. Outside air is needed for ventilation to provide thermal comfort as well as for diluting and removing indoor pollutants, odors, and moisture. Depending on the design and operational parameters of an HVAC system, the air supplied to the spaces can be entirely outdoor air (no recirculated air) or outdoor air mixed with indoor air drawn from the indoor spaces (return air). Using no recirculated air in a space requires more energy because large amounts of air must be conditioned for temperature and humidity levels when a mechanical system is used. In most cases using a percentage of return air mixed with the outside air is desirable for energy conservation. Increasing the amount of outside air in this air mixture to as high a level as is practical could potentially result in higher levels of human health, comfort, and productivity.

Ensuring that the air supplied is as clean as possible requires controlling the sources of pollutants and moisture within the ventilation system itself, cleaning the incoming outside air as much as possible prior to mixing it with the return air (most commonly this is going to be particulate filtration only, but where the outside air is very contaminated, gas-phase air filtration may also be used), effectively and continuously maintaining the hygiene of the HVAC system, and controlling indoor pollutant sources to minimize the spread of airborne pollutants. Additionally, a ventilation system should be capable of effectively distributing the ventilation air into occupied spaces and exhausting the return air from those spaces. Balancing the ventilation system for effective supply and exhaust rates is critical.

Many schools use unit-ventilator systems, a type of decentralized system, because their first costs (design and installation) are generally less than those of central systems: Unit ventilators eliminate the requirement for ducted supplies and returns (plenum or ducted). They distribute all the air from a single location, usually on the external wall of a room, thereby reducing ventilation effectiveness. They also typically do not meet the requirements for low ambient noise, necessary for acoustical quality associated with student learning (discussed in Chapter 6). Teachers often use the top of a unit ventilator as a storage shelf, so if this type of ventilation system is used in a green school, helping teachers understand the importance of not blocking air vents on the system is critical.

Central HVAC systems, which may supply small blocks of classrooms or entire sections of a school, require supply ducts and an air return system (plenum or ducted) to move the air to occupied spaces. Central systems can have multiple supply and return vents in a single classroom, potentially increasing ventilation effectiveness. Additionally, student comfort might improve since there is a lower probability of air blowing on students sitting on one side of a room.

The design of and materials used in the supply air ducts may have an influence on the long-term health and well-being of the students and on system maintainability. The noise of air moving in the ducts and its potential impact on student learning and teacher health are discussed in Chapter 6. Where the air has a high moisture content, the use of fiberglass-lined ductwork to attenuate noise transmission can support the growth of microbial contamination, if the system is not properly maintained.

The type of ventilation system used may depend on the climate. Throughout much of the United States, ventilation systems need to control humidity as well as temperature and ventilation rates throughout the year. This is particularly true where schools are used year-round. The ventilation system is the primary mechanism for indoor humidity control, particularly in hot and humid climates. Excess humidity in the ventilation system, ductwork, and the building spaces increases the probability of indoor microbial contamination. Active humidity control systems, such as desiccant systems, may be effective for controlling humidity through ventilation systems in hot and humid climates (Fischer and Bayer, 2003). Displacement ventilation is another form of active humidity control in cold climates (Melikov et al., 2005).

School buildings are intended to be used for many years, so it is critical that the ventilation system be designed to allow effective operations and maintenance practices. Sepännen et al. (2004b), in a literature review on the association of ventilation rate and human responses, reported that better hygiene, commissioning, operation and maintenance of air handling systems may be particularly important for reducing the negative effects of HVAC systems. Ventilation may also have harmful effects on indoor air quality and climate if not properly designed, installed, maintained, and operated. To be well-designed, HVAC systems should be easily accessible to facility maintenance staff for maintenance and repair activities.

Filtration

Indoor and outdoor particulates and certain VOCs can be effectively removed by filtration. Most filters are designed to collect particles larger than 10 μm but are relatively inefficient at removing submicron-sized particles. The location of the filters is critical and should ensure that both outside air and recirculated air are effectively filtered for particulate and VOCs removal before the airstream reaches occupants. In addition, filters should be located such that they can be consistently maintained. Efficient and effective filtration that removes particulate contamination to a level that protects building occupants and not just the equipment is essential (IOM, 2000, pp. 360-382). Particulate filtration having a Minimum Efficiency Reporting Value (MERV) of 11 or higher should be on all HVAC equipment supplying air to the occupied spaces of a building. Filters should fit snugly to prevent the bypass of air around the filter(s). The filters should be changed frequently and regularly to prevent them from becoming a source of indoor air pollution (Clausen, 2004; Hanssen, 2004). Additionally, filters should be kept dry, since wet filters may become microbially contaminated and thereby spread contamination throughout the area served by the ventilation system.

The deployment of electrostatic precipitators as a high-efficiency particulate filtration device may be desirable, particularly in schools near heavy traffic areas (IOM, 2000; Wargocki et al., 2005). If electrostatic air cleaners are used, they should be well maintained to minimize ozone formation. Additionally, indoor ozone sources should be controlled by proper maintenance of copiers and printers and exhausting these to the outside whenever possible. High ventilation rates will help to reduce the formation of by-products of indoor air chemical reactions. Ion-generating air cleaners can be sources of indoor air chemistry by-products (Wu and Lee, 2004; Bekö et al., 2006) and ozone and should be avoided.

Use of gaseous-phase filtration to remove gaseous pollutants from the supply air stream may be desirable in areas with significant amounts of outdoor air pollution. Gaseous-phase filters or filter media should be changed frequently and regularly.

Cleaning

Although to date no systematic research has examined the relationship of cleaning effectiveness to student and teacher health, student learning, or teacher productivity (Berry, 2005), a few studies have related methods for source reduction or control in schools to exposures to pollutants. Smedje and Nörback (2001) observed that classrooms with more frequent cleaning had lower concentrations of cat and dog antigen in settled dust. However, the study could not be repeated. Few studies have looked systematically at changes in exposure, health, or productivity in relation to changes in school building materials, cleaning products, or cleaning practices.

The literature on source reduction and control in homes, particularly those with asthmatic children, is more extensive (Takaro et al., 2004). Integrated pest management techniques have been shown to be effective in reducing antigen levels in homes (Phipatanakul et al., 2004) and have been shown to reduce pesticide levels in schools (Williams et al., 2005). However, whether they result in better health outcomes or improved productivity has not been determined (Phipatanakul et al., 2004; Williams et al., 2005).

The effects of air pollutants in schools can be reduced through proper design and maintenance practices for HVAC filters, drip pans, cooling coils and other elements. Simple measures such as closing windows during pollen season or prohibiting furry pets in a school may also be effective. In other cases, more subtle design considerations may be needed, for example, limiting food preparation, vending, and eating to certain areas with structural and surface finishes that allow for cleaning and easy pest control.

The choice of cleaning products and methods is also important. A study by Singer et al. (2006) shows that some cleaning products can yield compounds, including glycol ethers and terpenes, that can react with ozone to form a variety of secondary pollutants. Persons whose occupation is cleaning might encounter excessive exposures to those pollutants. Mitigation options include screening of product ingredients and increased ventilation before and after cleaning (Singer et al., 2006). Eliminating the use of air fresheners may also help to reduce the level of pollutants generated by indoor chemical reactions.

CURRENT GREEN SCHOOL GUIDELINES

Current green school guidelines contain many measures intended to improve indoor air quality. They typically endeavor to manage outdoor pollution in a number of ways. One is through the careful location of fresh air intakes to ensure that exhaust and other pollutants generated by trucks, buses, or cars are not fed back into the building; similarly, they emphasize locating air intakes away from loading docks and areas where standing water might pollute the fresh air being taken in. Anti-idling measures for cars, trucks, and buses focus on reducing CO and particulate pollution entering the air stream. The use of walk-off mats and grills, previously discussed under moisture management, is also intended to remove outside dust, dirt, and other pollutant sources that might be on the shoes of anyone entering a school.

Current green school guidelines also contain measures to protect and maintain clean ventilation pathways. They typically encourage the use of high-efficiency filters with a MERV of 10 or higher in all ventilation systems. Plenum air returns that can be contaminated by dust and microbial growth are to be replaced by ducted returns, but exposed fibrous insulation in supply or return ducts is discouraged unless there is double-wall construction to protect the insulation from airborne pollutants. The use of fibrous insulation in air ducts was vigorously debated by the committee because of the importance of mechanical noise abatement. Some green school guidelines discourage its use based on evidence that it absorbs VOCs and dust and encourages microbial growth inside duct linings that are difficult to clean. Effective noise reduction techniques for ducted HVAC systems and duct cleaning measures are both needed.

Finally, current green school guidelines contain numerous measures to reduce indoor pollutant sources: Two up-to-date industry standards are cited to ensure indoor air quality: ASHRAE 62.1-2004 IEQ (design guidelines) and the SMACNA IAQ Construction Guidelines. Green school measures include eliminating gas-fired pilot lights and discouraging fossil fuel burning equipment indoors to reduce the potential accumulation of exhaust fumes and the development of combustion products and particulate matter. Dedicated exhausts for all spaces that might contain chemicals—for example, storage rooms for cleaning equipment and supplies, photography laboratories, copy/print rooms, and vocational spaces—are encouraged.

Because construction materials create significant dust and outgassing contaminants that can remain in a building after completion, measures have been proposed to reduce indoor pollution from them. These include 72 hours of continuous ventilation of construction areas during the installation of materials that emit VOCs (ideally utilizing open windows and temporary fans rather than the HVAC system, which might absorb some of these contaminants); protecting supply and return ducts during construction; daily HEPA vacuuming for all soft surfaces such as carpets in or near construction areas; replacing all filters at the completion of construction; and providing for 28 days of continuous flushing of the building with outside air (except where high outdoor humidity could lead to mold growth) and then replacing the filters again. All of these measures are intended to reduce human exposure to higher levels of chemical emissions and dust/particulates in newly constructed areas.

Operable windows are typically suggested to allow for natural ventilation and decrease the demand for air conditioning. Such windows may provide a number of benefits: the ability to rapidly control temperature in an overheated classroom or in the event of an HVAC system failure; the ability to significantly increase the quantity of outside air ventilation to dissipate classroom pollution sources without resorting to the whole building HVAC; and the ability to locally ventilate classrooms undergoing renovation without using the whole building HVAC. The committee notes, however, there are concerns with operable windows that should be addressed, including the exposure of children to outside noise, pollution, and pollen, as well as the possible intrusion of rain and unwanted humidity.

To help ensure that longer term indoor air quality problems do not emerge, green school guidelines typically recommend the establishment of an indoor health and safety program for new or renovated schools, based on resources such as Environmental Protection Agency's Tools for Schools. This can help to establish operations and maintenance guidelines and clear communication channels so that indoor air quality problems can be prevented or identified and solved.

Despite emerging studies showing that thermal comfort may play a role in student performance, thermal comfort standards in green school guidelines are few. Typically they rely on compliance with the industry standard ASHRAE 55-2004 and do not address the control of humidity. In addition, they do not address radiant overheating by direct sunshine in the classroom in warm months or the potential contribution of solar heat through windows to offset heat losses in cool months. Managing sunshine on a seasonal basis is a critical aspect of ensuring thermal comfort.

FINDINGS AND RECOMMENDATIONS

Finding 4a: A robust body of scientific evidence indicates that the health of children and adults can be affected by indoor air quality. A growing body of evidence suggests that teacher productivity and student learning may also be affected by indoor air quality.

Finding 4b: Key factors in providing good indoor air quality are the ventilation rate; ventilation effectiveness; filter efficiency; the control of temperature, humidity, and excess moisture; and operations, maintenance, and cleaning practices.

Finding 4c: Indoor air pollutants and allergens from mold, pet dander, cockroaches, and rodents also contribute to increased respiratory and asthma symptoms among children and adults. Although limited data are available regarding exposure to these allergens in U.S. schools, studies in both school and nonschool environments support the notion that allergen levels can be decreased through good cleaning practices.

Finding 4d: The reduction of pollutant loads through increased ventilation and effective filtration has been shown to reduce the occurrence of building-associated symptoms (eye, nose, and throat irritations; headaches; fatigue; difficulty breathing; itching; and dry, irritated skin) and to improve the health and comfort of building occupants.

Finding 4e: There is evidence that ventilation rates in many schools do not meet current standards of the American Society for Heating, Refrigeration, and Air-Conditioning Engineers (ASHRAE). Available research indicates that increasing the ventilation rate to exceed the current ASHRAE standard will further improve comfort and productivity.

Finding 4f: Scientific evidence indicates that increased ventilation rates can reduce the incidence of building-related symptoms, reduce pollutant loads associated with asthma and other respiratory diseases, and improve the productivity of adult workers. Increased ventilation rates may also reduce the potential for reactions among airborne pollutants that generate irritating products and may improve perceived air quality. However, the research conducted to date has not established an upper limit on the ventilation rates, above which the benefits of outside air begin to decline.

Finding 4g: Research comparing the effects of natural versus mechanical ventilation on human health is inconclusive. However, there is evidence that improper design, maintenance, and operation of mechanical ventilation systems contribute to adverse health effects, including building-related symptoms among occupants.

Finding 4h: Studies in office buildings indicate that productivity declines if room temperatures are too high. However, there are few studies investigating the relationship of room temperatures to student learning, teacher productivity, and occupant comfort.

Finding 4i: To date, no systematic research has examined the relationship of cleaning effectiveness to student and teacher health, student learning, or teacher productivity. Few studies have looked systematically at changes in exposures, health, or productivity based on changes in building materials, cleaning products, or cleaning practices.

Recommendation 4a: Future green school guidelines should ensure that, as a minimum, ventilation rates in schools meet current ASHRAE standards overall and as they relate to specific spaces. Future guidelines should also give consideration to planning for ventilation systems that can be easily adapted to meet evolving standards for ventilation rates, temperature, and humidity control.

Recommendation 4b: Future green school guidelines should emphasize the importance of appropriate operation and preventive maintenance practices for ventilation systems, including replacing filters, cleaning coils and drip pans to prevent them from becoming a source of air pollution, microbial contamination, and mold growth. These systems should be designed to allow easy access for maintenance and repair. The Environmental Protection Agency's Tools for Schools program is a well-recognized source of information on methods for achieving good indoor air quality.

Recommendation 4c: Additional research should be conducted to document the full range of costs and benefits of ventilation rates that exceed the current ASHRAE standard and to determine optimum temperature ranges for supporting student learning, teacher productivity, and occupant comfort in school buildings.

Recommendation 4d: Studies should be conducted to examine the relationships of exposures from building materials, cleaning products, and cleaning effectiveness to student and teacher health, student learning, and teacher productivity.

5

Lighting and Human Performance

For purposes of learning, performance, and productivity, lighting in a school building should allow people to see to read, to see others with whom they are communicating, and to perform other visual tasks associated with learning, teaching, and school administration. Lighting can be provided by electric systems or by daylight through windows, clerestories, and skylights. Typically, school buildings use a combination of electric lighting and daylight.

When evaluating the performance of any lighting system, electric or daylight, its impact on two biological systems—the visual and the circadian—needs to be considered, together with the physical attributes of light that differentially affect these systems (Figure 5.1).

LIGHTING FOR VISUAL PERFORMANCE

The visual system functions as a very quick remote-sensing mechanism that alerts us to environmental changes and enables us to identify nearby threats and opportunities. The visual system is fairly well understood for adult populations in regard to the effects of light on both appearance (what things look like) and visual performance (how well visual information is processed). For example, there is a complete model of visual performance available for predicting the impact of background luminance (light level), target contrast, target size, and observer age from 18 to 65 years (Rea and Ouellette, 1991). Presumably, most school-age students should behave like 18-year-olds in regard to visual performance, but this has not been systematically studied. In general, given the characteristics

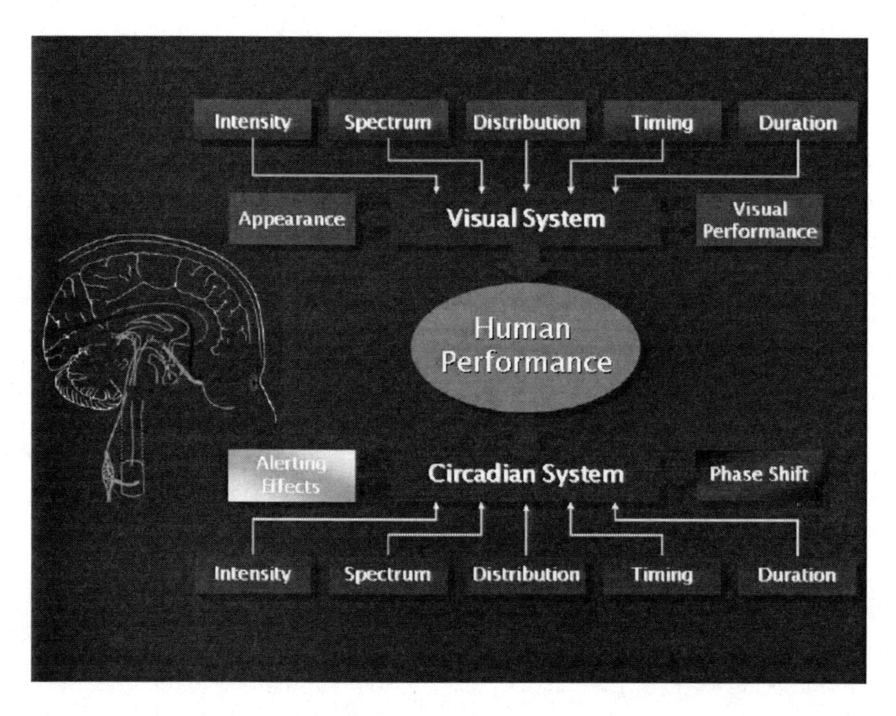

FIGURE 5.1 Light as it affects human performance and health through the visual and circadian systems. SOURCE: Lighting Research Center.

of the visual response functions, it can probably be concluded that most lighting and task conditions are adequate for most students. Going the next step, however, is more tenuous because there is no evidence that the quality or quantity of light directly affects student learning (Larson, 1965; Demos et al., 1967; Boyce et al., 2003).

For a large majority of the people working in buildings, lighting for vision during the day is quite adequate, in part because people have very flexible visual systems and adjust their posture in response to the available lighting conditions. For example, the dimmer the light, the closer one holds the reading materials to maintain a constant ability to read. Laboratory experiments show that young people with normal eyesight will systematically adjust the eye-to-task spacing to maintain good task visibility, either by moving closer to the visual task or shifting posture to avoid reflected glare (Rea et al., 1985). A flexible visual system combined with a flexible body provides most people with the ability to adapt to less than ideal lighting environments.

A significant minority of the school-age population may not have properly corrected eyesight, however, and may not be able to take full advantage of adaptive strategies to see educational materials in the classroom. In a large study using the 1996-1997 National Health Interview Survey, Kemper et al. (2004) determined that approximately 25 percent of U.S. school-age children have corrective lenses (eyeglasses or contact lenses). They showed that the prevalence of corrective lenses is related to several population factors such as age (older children are more likely to wear corrective lenses), ethnicity (black and Hispanic children are less likely to wear corrective lenses), income (poorer children are less likely to wear corrective lenses), and gender (girls are more likely to wear corrective lenses than boys). Of particular note, insurance coverage appears to be a major factor in the prevalence of corrective lenses in school-age children. The actual percentage of corrective lens wearers is probably smaller than the percentage of children who need them. Moreover, it is unknown whether students' corrective lenses actually provide the proper refraction. Because lighting and task variables as well as proper corrective lenses determine visual performance, it seems likely that at least some students are not able to adequately see educational materials in the classroom.

Within the context of the school environment, then, what constitutes adequate lighting for the majority of the students may be inadequate for others. It could be hypothesized that for those students who need but do not have corrective lenses, supplemental daylight may offer a significant advantage by providing higher light levels and better distribution of light (side light) to minimize shadows than would be provided by electric lighting alone. However, the potential for daylight in classrooms to improve the visual performance of children with or without properly corrected eyesight has not been systematically studied.

Some portion of the adult population (teachers, administrators, support staff) will likely also have visual problems. Normal aging involves a continuous loss of visual accommodative ability, known as presbyopia, from about 20 years of age to about 65 (Weale, 1992). Until about age 45 the gradual loss in ability to focus on close objects is hardly noticed, but after this age nearly everyone begins to adopt new strategies to see small objects. Instead of getting closer to an object to see it as they did when they were younger, people with presbyopia actually move the object farther away from their eyes, or they place it under a bright light, usually provided by a window or skylight. Eventually, everyone needs optical aids, such as bifocals, reading glasses, or contact lenses to read normal print, but the use of bright light from the right direction will continue to be a strategy employed by older people to see small objects. Thus, the visual performance of adults older than 45 in a school building is likely to be negatively affected by poor lighting.

Glare and Visual Performance

There is much less certainty in predicting visual comfort in regard to glare, even in adults, than in predicting visual performance (Rea, 2000; Boyce, 2003). A clear distinction is made between glare that reduces visual performance (disability glare) and glare that does *not* (discomfort glare). Disability glare can be precisely predicted for a given individual and for the general population. As would be expected, disability glare becomes more problematic with age owing to physical changes of the eye, particularly from the scattering of light within the eye by the crystalline lens (Weale, 1963).

Formulas exist for calculating discomfort glare, and these are often used to characterize the lighting layout for a space using commonly available lighting software. However, collective understanding of the mechanisms underlying discomfort glare is rather poor (Boyce, 2003). Psychological phenomena appear to contribute significantly to discomfort glare, so that, for example, bright flashing lights in an office are highly uncomfortable, but the same lights can be highly desirable in a nightclub for dancing. Therefore, although disability glare is the same in both applications, discomfort glare is not. In this context, and given the strong psychological component to discomfort glare, predicting visual comfort in school-age children is an area that requires more research.

Daylighting and Student Achievement

Several studies investigating the effect of daylighting on student achievement were conducted by the Heschong-Mahone Group between 1999 and 2003. In the 1999 study, data were obtained from elementary school districts in three locations: Orange County, California; Seattle, Washington; and Fort Collins, Colorado (Heschong-Mahone Group, 1999). The study looked for a correlation between the amount of daylight provided by a student's classroom environment and test scores. Test results for more than 21,000 students in these districts were analyzed. Demographic data sets, architectural plans, aerial photographs, the presence of skylights, maintenance records, and daylighting conditions for more than 2,000 classrooms were among the factors reviewed.

The study developed a regression model for approximately 150 independent variables (e.g., teacher salaries, grade level, attendance), including available daylight, which was represented by one of five different levels, or "daylight codes." Although the regression analysis leads to a prediction that an increase in the value of the daylight code can increase scores in both math and English by more than 20 percent, a closer examination of the results shows that only 0.3 percent of the variance in the regression model is explained by daylight code (Boyce, 2004). This is a very small effect and one that cannot be considered statistically significant. As noted below, these results could not be replicated in a subsequent study.

A reanalysis of the Orange County and Seattle data was undertaken in 2001 to evaluate additional variables that might have a confounding influence, including teacher assignments (Heschong-Mahone Group, 2001). In 2003 a third study was undertaken to see whether the original methodology and findings would hold up when data came from a school district (Fresno, California) with a different climate and curriculum. The preliminary statistical analyses used the same models as the previous studies. In the Fresno study, the holistic variable "daylight code" was found to be "not significant in predicting student performance. It had the least explanatory power of the variables considered, and the lowest significance level" (Heschong-Mahone Group, 2003, p. viii).

The authors proceeded with more detailed multilinear regression (statistical) analysis to see whether they could gain some insight into why the daylight code was not significant in Fresno as it had been in the earlier studies. Among the authors' conclusions were that sources of glare negatively affect student learning; direct sun penetration into classrooms, especially through unshaded east- or south-facing windows, negatively affects student performance, likely causing both glare and thermal discomfort; blinds or curtains allow teachers to control the intermittent sources of glare or visual distraction through the windows; absence of teacher control of windows negatively affects student performance (Heschong-Mahone Group, 2003, p. ix). They summarized their results as follows:

> Characteristics describing windows were generally quite stable in their association with better or worse student performance. Variables describing a better view out of windows always entered the equations as positive and highly significant, while variables describing glare, sun penetration, and lack of visual control always entered the models as negative (Heschong-Mahone Group, 2003, p. viii).

Because of the inconsistent results of this limited number of studies, there is insufficient evidence at this time to determine whether or not an association exists between daylight and student learning.

Views and Performance

Windows can be a major source of light in school buildings. Despite the greater thermal energy losses and higher initial costs they incur relative to insulated walls, people universally prefer having windows. First and foremost, windows provide a view to the outside. Depending on the context, views can be beneficial or distracting. Windows can also provide high light levels, and, when properly located, they can provide ideal distribution of light for performing visual tasks that do not involve self-luminous displays, such as computer screens, or audiovisual presentations. Skylights and clerestories can also be good sources of illumination even though they do not provide a view.

Studies conducted using subjective ratings from adults show that people like views from windows (Markus, 1965; Jackson and Holmes, 1973; Ne'eman and Longmore, 1973; Collins, 1975; Ludlow, 1976; Cuttle, 1983; Heerwagen and Heerwagen, 1986; Leslie and Hartleb, 1990, 1991; Boubekri et al., 1991) and that architectural spaces with views from windows command higher prices (Boyce et al., 2003). Although these findings for adult populations are interesting, the relationship between view and performance of students remains largely undocumented, with the exception of one study that attempted to relate children's hormone levels to behavior in classrooms with and without windows.

Kuller and Lindsten (1992) studied children's health and behavior in classrooms with and without windows for an entire academic year. They concluded that work in classrooms without windows affected the basic pattern of the hormone cortisol, which is associated with stress, and could therefore have a negative effect on children's health and concentration. This finding is strictly suggestive, however, because no direct relationship between cortisol levels and student performance and health was established (Rusak et al., 1997).

LIGHTING AND THE CIRCADIAN SYSTEM

The circadian system involves biological rhythms that repeat at approximately 24-hour intervals. The behavior of all terrestrial species, including humans, is driven by an internal clock synchronized to the solar light-dark cycle. Indeed, light is the primary stimulus for the internal clock. The circadian system regulates not only overt patterns of behavior such as activity and rest, but also bodily function at the cellular level, such as the cell cycle (Moore, 1997).

Current lighting technologies and lighting standards are designed exclusively for providing visual sensation. However, light affects the visual system very differently than it affects the circadian system. Relative to the visual system that underlies conventional photometry and all lighting standards, the circadian system needs a much higher light level on the retina for activation (McIntyre et al., 1989a,b); it has a peak spectral sensitivity to much shorter wavelengths (Brainard et al., 2001; Thapan et al., 2001); it has greater sensitivity to light in the inferior retina (such as would be involved when a person looks at the sky) than in the superior retina (Glickman et al., 2003); it requires much longer exposures for activation (McIntyre et al., 1989a,b; Rea et al., 2002); and, most important, its sensitivity to light depends on the time of day (Jewett et al., 1997).

There is a growing body of literature indicating that the effect of light on circadian rhythms can affect productivity as well as health. Seasonal affective disorder (SAD), or the "winter blues," is recognized by the medical community as a psychiatric disorder. Apparently, seasonal reductions in the amount of daylight available in the winter at extreme northern and southern latitudes can induce depression (Rosenthal, 1998). Light treatment, typically provided as bright light from electric lighting systems, is recognized by the medical community as the preferred method of treating SAD (Rosenthal et al., 1985).

The incidence of SAD in school-age children is poorly documented, although it has been reported that adults who experience SAD also experienced it as a child. It seems, too, that postpubescent young women are more likely to experience SAD (Rosenthal, 1998). Depending on latitude, between 4 and 30 percent of the adult population, usually women, experience some symptoms of seasonal depression (Rosenthal, 1998), and this might also be true for teachers. Less learning might be expected on the part of children who experience symptoms of seasonal depression, so lighting might play a very important role in the design of a green school at northern latitudes. Systematic attempts to alleviate seasonal depression in children through lighting design have not been undertaken, but these early findings suggest a reconsideration of the role that light, particularly daylight, plays in the classroom.

Nearly half the population experiences some form of sleep disorder (National Sleep Foundation, 2005). Poor sleep directly affects a person's ability to perform tasks and learn new tasks (Jennings et al., 2003; Heuer et al., 2004). Light and dark have a dramatic impact on sleep quality (Turek and Zee, 1999; Reid and Zee, 2004). Adolescents in particular commonly go to sleep late (after midnight) and have difficulty getting up early (before 7:00 a.m.) to go to school (Carskadon et al., 1998). In extreme cases this difficulty in falling asleep late and getting up early is diagnosed as delayed sleep phase syndrome (DSPS). Many school age children with DSPS must get special training or even repeat grades because of poor attendance and performance. Light is a recognized treatment for this disorder, and a regular light-dark cycle may have broader implications for sleep quality in a larger group of children.

Recent research at the other end of the age spectrum shows that light treatment can consolidate sleep and increase sleep efficiency during the night in older people (Satlin et al., 1992; Fetveit et al., 2003; van Someren et al., 1997; Figueiro and Rea, 2005). A regimen of bright light at school during the day, together with dark nights at home, may increase the attendance and performance of school age children; however, there have been no systematic studies for this age group.

SOLUTIONS/DESIGN REQUIREMENTS FOR VISUAL PERFORMANCE

Sources of Light

Electric lighting systems have a number of components: luminaires, lamps (incandescent, fluorescent, high-intensity discharge [HID]), ballasts (except when using incandescent lamps), and controls. Electric lighting systems differ in the amount of power they require to operate and in the amount and direction of light they are able to generate to meet design objectives. They also vary in their initial cost, ease of maintenance and commissioning, and expected life. Also important, electric lighting systems vary in their ability to provide good color rendering and low levels of glare, flicker, and noise.

Fluorescent lighting systems are the most prevalent sources of general illumination in schools. Modern fluorescent systems (T8 and T5 lamps with electronic ballasts) can provide low cost, long life, high efficacy, good color, low levels of noise, and flicker. Other sources of illumination, including incandescent and HID lamps, can be specified to best accomplish specific design objectives, from outdoor applications such as sports fields to illuminating pictures or works of art (Rea and Bullough, 2001).

Windows are an important part of a school's design as they relate to lighting. They allow for high light levels and, when properly located, ideal lighting configurations for visual tasks not involving self-luminous displays, such as computer screens, or audiovisual presentations. Windows are also the largest sources of glare in a classroom. However, glare can be controlled with fixed overhangs and blinds or window treatments that can be manually operated. Methods to control light from skylights and clerestories are also needed because the distribution and level of light changes as the day progresses.

A key difference between electric light and daylight is that electric light is almost always static, whereas daylight is ever-changing over the course of a day, with weather conditions, and with season. Daylight will also be different from one school to another, depending on building orientation and site, climate, and latitude, so that a cookie-cutter building design will rarely provide ideal lighting.

The dynamic nature of daylight, together with the wide range of intensities and distributions, demands a sophisticated understanding of its interactions with a building and the building's spaces: A much more sophisticated understanding is required for using daylight effectively than for using electric lighting effectively in school design. In some circumstances it may be desirable to conduct detailed lighting, heating, and cooling simulations in order to gain such an understanding.

Lighting Criteria for Schools

Light levels in school buildings are strongly influenced by the expected visual performance requirements for a given task. In general, higher illuminance levels are recommended for specialized tasks such as reading and writing than for less demanding visual tasks such as eating or walking. Lower illuminance levels are also recommended for public spaces where reading and visual inspection are only occasionally performed or where there is no time pressure to complete the task. For these reasons, lighting should be designed not just with respect to the source of illumination or the individual components needed to create the entire lighting system but should instead be designed with respect to the integrated system of enclosure design and controls, space configurations and surface finishes, and fixture components, all of them in relation to the task requirements: In schools, it is inappropriate to require specific types of luminaires and lamps without consideration of the space layout. This is true for new construction, significant renovation, or retrofit of school buildings.

Illumination on horizontal work surfaces from any type of electric luminaire will be equally satisfactory with regard to occupant visual performance. Among the recommended illuminance levels are these:

- Desks (300-500 lux on a desktop),
- Chalkboard (500 lux on a vertical surface),
- Corridors (100 lux on the floor), and
- Art rooms (500 lux on a desktop, 300 lux in the vertical plane) (Rea, 2000).

Lighting for chalkboards and whiteboards should be directed from a fixture to the specific surface to be illuminated (rather than coming from general illumination). Application efficacy (lumens per solid angle of the surface area per watt) is the correct photometric for determining the most energy efficient lighting for these uses (Rea and Bullough, 2001; Bullough and Rea, 2004).

Color rendering index (CRI) is commonly used by the industry as a measure of a lamp's ability to make objects appear "natural." A CRI of 80 for fluorescent lighting systems is probably appropriate, although a CRI as low as 70 can also provide satisfactory color rendering in most cases. It should be pointed out, however, that as new lighting technologies (e.g., LEDs) become more prevalent, CRI may not be a useful measure of source color rendering. Indeed, industry and government are presently examining the utility of CRI for LEDs. In the meantime, CRI should be supplemented with two other measures of color rendering—gamut area (GA), which is related to the saturation or vividness of hues, and full spectrum color index (FSCI), which is related to fine color discrimination. All three measures are useful for specifying lighting sources (Rea et al., 2004).

Various lamp and luminaire combinations can be equally effective at producing energy-efficient, long-life, low-glare lighting systems. As noted previously, windows can be the main source of glare in classrooms. The easiest and most cost-effective method for controlling glare from windows is to provide manual blinds or other window treatments that can be adjusted by teachers as the need arises. The current technologies for automatic "daylight harvesting"[1] require specialized expertise to design, operate, and maintain them. Although some manufacturers will be able to ensure proper performance, these systems are invariably expensive and can rarely be justified economically (Bullough and Wolsey, 1998; Bierman and Conway, 2000).

CURRENT GREEN SCHOOL GUIDELINES

Current guidelines for green schools usually focus on energy-efficient lighting technologies and components and the use of daylight to further conserve energy when addressing lighting requirements. They do not give guidance for lighting design that supports the visual performance of children and adults in various tasks and for different school room configurations, layouts, and surface finishes. Excellent resources for developing such guidance in the future include the current consensus-based lighting design guidelines from the Illuminating Engineering Society of North America.

FINDINGS AND RECOMMENDATIONS

Finding 5a: The research findings from studies of adult populations seem to indicate clearly that the visual conditions in schools resulting from both electric lighting and natural light (daylighting) should be adequate for most children and adults, although this supposition cannot be supported by direct evidence.

[1]"Daylight harvesting" involves using photo sensors to detect daylight levels and automatically adjust the output level of electric lighting to create a balance. The goal is energy savings.

Finding 5b: There is concern that a significant percentage of students in classrooms do not have properly corrected eyesight, and so the general lighting conditions suitable for visual functioning by most students may be inadequate for those students who need but do not have corrective lenses. It could be hypothesized that daylight might benefit these children by providing higher light levels and better light distribution (side light) than would electric lighting alone. However, the potential advantages of daylight in classrooms for improving the visual performance of children without properly corrected eyesight has not been systematically studied.

Finding 5c: Current green school guidelines typically focus on energy-efficient lighting technologies and components and the use of daylight to further conserve energy when addressing lighting requirements. Guidance for lighting design that supports the visual performance of children and adults, based on task, school room configurations, layout, and surface finishes, is not provided.

Finding 5d: Windows and clerestories can supplement electric light sources, providing high light levels, and good color rendering. Light from these sources is ever-changing and can cause glare unless appropriately managed. Currently, there is insufficient scientific evidence to determine whether or not an association exists between daylight and student achievement.

Finding 5e: A growing body of evidence suggests that lighting may play an important nonvisual role in human health and well-being through the circadian system. However, little is known about the effects of lighting in schools on student achievement or health through the circadian system.

Recommendation 5a: Future green school guidelines should seek to support the visual performance of students, teachers, and other adults by encouraging the design of lighting systems based on task, school room configurations, layout, and surface finishes. Lighting system performance should be evaluated in its entirety, not solely on the source of illumination or on individual components.

Recommendation 5b: Future green school guidelines for the design and application of electric lighting systems should conform to the latest published engineering practices, such as the consensus lighting recommendations of the Illuminating Engineering Society of North America.

Recommendation 5c: Green school guidelines that encourage the extensive use of daylight should address electric control systems and specify easily operated manual blinds or other types of window treatments to control excessive sunlight or glare.

Recommendation 5d: Because light is important in regulating daily biological cycles, both acute effects on learning and lifelong effects on children's health should be researched, particularly the role that lighting in school environments plays in regulating sleep and wakefulness in children.

6

Acoustical Quality, Student Learning, and Teacher Health

Good acoustical quality in classrooms is critical for student learning. Research has shown that noise exposure affects educational outcomes and provides evidence of mechanisms that explain the effects of noise on learning. Speech intelligibility studies indicate that students' ability to recognize speech sounds is decreased by even modest levels of ambient noise, and this effect is magnified for younger children. This problem is frequently not appreciated by adults, who are better able to recognize speech in the presence of noise.

Most learning activities in school classrooms, especially for younger children, involve speaking and listening as the primary communication modes: Students learn by listening to the teacher and to each other (Goodland, 1983). Excessive background noise or reverberation (i.e., multiple delayed reflections of the original sound) can interfere with speech perception and thereby impair learning. Careful attention to acoustical design requirements, then, is essential for creating an effective learning environment. Nonetheless, a 1995 report of the U.S. General Accounting Office estimated that the acoustical quality in approximately 22,000 U.S. schools attended by 11 million students was unsatisfactory (GAO, 1995a).

Figure 6.1 illustrates the four typical major sources of noise in classrooms and reflected speech sounds (reverberation).

HVAC systems are perhaps the most common source of ambient noise in classrooms. However, significant levels of noise can also be transmitted through walls and windows from outdoors or from adjacent indoor spaces. Reflected sound within a classroom, if uncontrolled, degrades speech intelligibility.

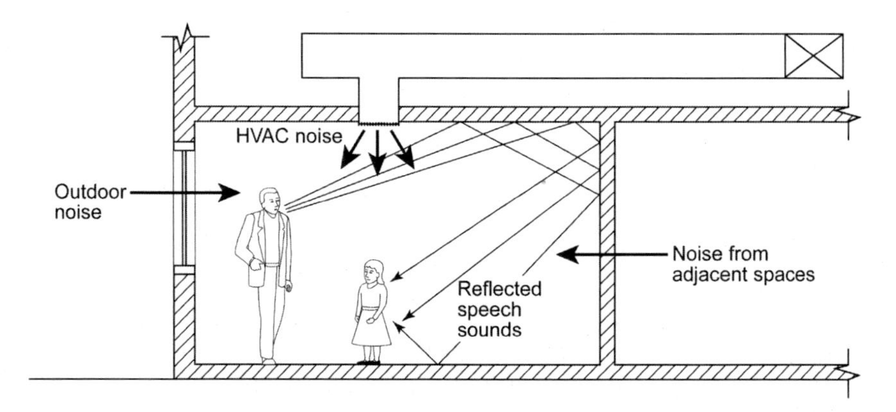

FIGURE 6.1 Key noise and reverberation sources in classrooms.

One way to describe the desired acoustical quality in a classroom is to specify an acceptable maximum ambient noise level. This level is measured in terms of A-weighted sound levels or octave band sound levels that can be used to determine other measures such as noise criterion (NC), room criterion (RC), or balanced noise criterion (BNC) values. By combining the effects of sound at different frequencies in a manner similar to that which takes place in the human hearing system, these measures rate the loudness of sound to listeners. A second way to describe acceptable room acoustical quality is to specify the reverberation time, which is approximately the time it takes for a loud sound to die away to inaudibility after the source is turned off. Reverberation times increase with room volume and decrease as sound-absorbing material is added to a room. However, excessive sound-absorbing treatments will have the negative effect of reducing speech levels and degrading the intelligibility of speech in a classroom.

People's ability to understand speech is influenced largely by how loud speech sounds are relative to ambient noise or any other competing sounds, hence the importance of an adequate signal-to-noise ratio (i.e., speech to background noise ratio) for a classroom to function well. Reverberant sound causes one word to smear into the next and can decrease the intelligibility of speech. Acoustical design should be aimed at improving the recognition of speech sounds in the classroom. The focus should be first on reducing unwanted noise and then on controlling excessive reverberation. Good acoustical design can facilitate learning by allowing for more accurate verbal interaction and less repetition among teachers and students because spoken words are clearly understood. There is also evidence that good acoustical design may have a health benefit for teachers by reducing the potential for vocal impairment, and it may have the ancillary benefit of reducing teacher absenteeism. This issue is discussed later in this chapter.

EFFECTS OF EXCESSIVE NOISE

Excessive noise can interfere with learning by affecting memory (Hygge, 2003) and acting as a distraction that impairs a student's ability to pay attention. The ability to pay attention is most important when students are engaged in tasks that demand higher mental processes, such as learning new concepts, or when teachers are presenting new or complex information (Hartman, 1946). (See also Anderson, 2004, for a review of the effects of noise on children and classroom acoustics issues.)

Excessive background noise in a classroom can come from outside the building (aircraft and traffic noise, lawn mowers and other equipment, or students engaging in sports activities) or from within it (heating, ventilation, air conditioning, plumbing systems, adjacent classrooms, hallways, gymnasiums, or music rooms) or from the students themselves. The level of residual noise from the students may be dominant, but is strongly related to the ambient noise in the room. That is, student chatter will increase as the general level of ambient noise increases, an example of the Lombard effect (Junqua, 1996).[1] Thus, it is important to minimize all other sources of noise to ensure lower levels of student noise. It is equally important to educate teachers about the effects of noise on speech communication. As adults, teachers may not appreciate the additional problems that noise creates for younger listeners.

Although the importance of classroom acoustics to educational outcomes is well supported in the research literature, it is frequently ignored by school officials and by those designing schools. Anderson (2004, p. 118) suggests there are at least four reasons for this:

> First, administrators walk into classrooms, listen briefly with adult ears and do not recognize that auditory immaturity causes young children to experience greater listening problems and less coping ability than the mature auditory systems of adults. Second, most school administrators have not been exposed to the extensive body of research that illustrates the effects of excessive background noise and/or reverberation on students' listening, learning, and behaviour. Third, administrators often believe that good classroom acoustics are only needed for children with hearing impairment and that children with hearing loss and auditory learning difficulties comprise only a very small proportion of the children educated in inclusive classroom settings. Fourth, school administra-

[1]The Lombard effect is the tendency to increase one's vocal intensity in noisy situations.

tors are typically unaware of the health issues faced by teachers who instruct in noisy classrooms and the expense that this may cost the school district.

Effects of Noise and Reverberation on Speech Perception

Speech perception studies have investigated how interference from noise and reverberation influences the recognition of syllables, words, or sentences in classrooms. Kindergarten and first and second grades are the main years in which children learn to break written words into their phonetic components and acquire the ability to read. Careful listening is needed to develop the ability to discriminate among minor differences in words such as pet, pit, pot, put, and pat (Anderson, 2004). Such differences can be lost in a noisy environment, so young children require the higher signal-to-noise ratios provided by quieter conditions.

The impacts of excessive noise vary according to the age of the students, because the ability to focus on speech is a developmental skill that evolves and does not mature until ages 13 to 15 years. As children mature they tend to develop strategies to cope with noise levels. Accordingly, young children require better acoustical environments than do adult listeners to achieve equivalent word recognition scores (Elliott et al., 1979; Elliott, 1979; Neuman and Hochberg, 1983). Classrooms of younger children are also found to be noisier (Picard and Bradley, 2001).

A student's difficulty in understanding speech in noisy situations may not be recognized by teachers, building designers, or other adults. That is, adults cannot rely on their own perception of speech under adverse listening conditions to recognize a child's difficulty under the same conditions. Elliott et al. (1979) found that the ability to recognize sentences in noisy environments improves systematically with age for children who are 5, 6, 7, 8, and 10 years old. Similar effects of children's age on speech perception in noisy environments were reported by Finitzo-Hieber and Tillman (1978) and Marshall (1987).

Although it is clear that children need quieter conditions than adults to achieve high speech recognition scores (Elliott, 1979) and that the younger the children, the quieter the conditions should be, the results of the various studies are very different and do not agree with the results of similar tests carried out in actual classrooms (Bradley and Sato, 2004). This is thought to be so because the early laboratory studies used monaural (one-ear) headphones to present speech and noise signals, which increases the negative effects of noise and reverberation. Listening naturally with two ears confers a binaural advantage, enabling higher speech recognition scores in noisy environments. Nábělek and Robinson (1982) found binaural advantages of a similar magnitude for 10-year-olds and adults. However,

Neuman and Hochberg (1983) found greater binaural advantages for the youngest children (5-year-olds). That is, the speech recognition scores of the 5-year-olds improved when they listened normally with two ears rather than monaurally. It is clear that the older studies in which monaural headphones were used exaggerated the negative effects of noise on speech recognition. Nonetheless, it is well established that children need quieter conditions than adults to understand speech well and that their ability to understand speech in everyday noise improves with age.

There are also studies suggesting that the negative effects of excessive reverberation are more acute for younger listeners (Nábělek and Pickett, 1974; Nábělek and Robinson, 1982; Neuman and Hochberg, 1983). Some early studies used monaural headphone presentation of the test signals (Finitzo-Hieber and Tillman, 1978), which would exaggerate the negative effects of reverberation on speech recognition scores (Moncur and Dirks, 1967). Because shorter reverberation times led to improved speech recognition scores in these tests, they caused some to recommend very short reverberation times for classrooms. However, such conclusions are based on an incomplete understanding of room acoustics in that it is not possible to have both increased speech level (to maximize signal-to-noise) and reduced reverberation time. For example, adding sound-absorbing material to a room will reduce reverberation times (which usually helps speech intelligibility) but at the same time will reduce speech levels, leading to reduced signal-to-noise ratios and reduced intelligibility scores. Although reverberant speech sound degrades the intelligibility of speech, the early arriving reflections of the speech sound (those that arrive at the listener within about 0.05 seconds after the direct sound) significantly enhance the apparent loudness of the speech and improve speech intelligibility (Bradley et al., 2003). In many situations, it is the early reflection energy that makes it possible to understand speech in rooms (e.g., children listening farther from the teacher or children listening while the teacher's voice is directed away from them). Thus adding too much absorptive material to reduce reverberation time will reduce speech loudness and impair intelligibility. This is why acoustics textbooks have for many decades referred to the need for optimum rather than minimum reverberation times (Knudsen and Harris, 1950).

Because too much noise and the reduced signal-to-noise ratios that accompany it are a more or less ubiquitous problem in classrooms, there must be enough reflected speech sound to maintain adequate loudness. Excessive noise is typically a more significant problem than is too much reverberation. Noise levels in classrooms can easily be 10 dB (or much more) too loud, indicating 10 times too much noise energy. In contrast, it is almost impossible for reverberation times to be 10 times too long. An optimum reverberation time will ensure that there are adequate speech levels without excessive reverberation. A few studies have looked at this problem (e.g., Reich and Bradley, 1998; Hodgson and Nosal, 2002), but none have focused on the particular needs of young children. Although these studies are not conclusive, they indicate that a reverberation time in the range close to conventional recommendations (0.4 to 0.7 s) is probably acceptable. Following American National Standards Institute (ANSI) Standard 12.6, "Acoustical Performance Criteria, Design Requirements and Guidelines for Schools," which recommends designing for a reverberation time of 0.6 s or a little less, and appreciating that a small deviation from this probably does not have a large effect on speech recognition scores in classrooms, seems to be the best advice now available.

EXCESSIVE NOISE AND STUDENT ACHIEVEMENT

In addition to degrading children's ability to recognize speech sounds, excessive noise can also interfere with the performance of various tasks. For example, children have been found to be more likely to give up on solving difficult puzzles in noisier situations (Cohen et al., 1980). Interfering noise can be more distracting for younger children (Higgins and Turnure, 1984), and speech sounds can be more disturbing than neutral, ventilation-type noises (Carhart et al., 1969; Elliott et al., 1979). This latter result suggests that speech sounds from adjacent classrooms are much more distracting than many other types of noise. The interference with the recognition of speech sounds and with various tasks may explain, in part, reported impaired student achievement in noisy environments. Green school guidelines should address the design of HVAC systems and walls and doors separating classrooms and corridors and the acoustic quality of windows and walls to the outdoors.

Transportation Noise Sources

The most substantial body of research related to noise and student performance in the classroom examines the impacts of noise from road traffic, trains, and aircraft. Since the 1970s, a number of studies have been conducted that compare the reading skills of students in schools exposed to transportation noise with the reading skills of students in schools in quieter areas. A study in the early 1970s looked at the performance of children in a New York school that was next to the tracks of an elevated train. Over a 3-year period, the aggregate scores of students in grades two, four, and six on the train side of the school were compared with those of students on the nontrain side of the school. Students on the noisy side lagged 3 to 4 months behind in reading compared with students on the quieter side. After the train tracks were treated to abate the noise, reading levels of children on that side of the building improved (Bronzaft and McCarthy, 1975; Bronzaft, 1981).

A 1982 study of students in New York schools under and not under flight paths matched the students for socioeconomic status, race, gender, hearing loss, mother's education level, and English as a second language (Green et al., 1982). The study found that high levels of environmental noise were inversely related to reading ability in elementary school children. The authors were also able to conclude that the reading deficits were due to chronic noise exposure and not to the noise levels at the time of the test. A later study of students in New York matched students/schools for low socioeconomic status, student absenteeism, and teacher experience and then analyzed reading achievement test scores for grades two through six (Evans and Maxwell, 1997). The analysis found that a higher percentage of students in noisy schools were reading 1 to 2 years below their grade level.

A study of schools near Munich airport looked at the cognitive effects on children before and after the airport was moved to a new location. The study found impaired reading comprehension in third- and fourth-grade children before the move. Children attending schools near the old airport had significantly more errors on a standardized reading test than students from quieter communities. Further, reading comprehension deteriorated in children in schools near the new airport (Hygge et al., 1996).

A 1997 study comparing students from two schools near Heathrow Airport found a significant association between noise and reading comprehension that could not be accounted for by annoyance, social class, or other factors (Haines et al., 2001a,b).

In one of the most comprehensive and rigorous studies to date, Stansfeld et al. (2005) conducted a cross-national, cross-sectional study to assess the effect of exposure to aircraft and road traffic noise on cognitive performance (reading comprehension) and health in children. The study assessed 2,844 children ages 9 and 10 in 89 schools in the United Kingdom, Spain, and the Netherlands in 2002. Schools in all three countries were selected to represent varying levels of exposure to aircraft and traffic noise. The selected schools were matched by students' socioeconomic status, the primary language spoken at home, and other factors. External noise was measured in decibels, and reading comprehension was assessed using standardized and normalized tests routinely used in each country.

Tests were also conducted to measure students' recognition and recall (episodic memory), sustained attention, working memory, and prospective memory. Socioeconomic characteristics were assessed as potential confounding factors, and pilot studies were conducted beforehand to assess the feasibility, reliability, validity, and psychometric properties of the cognitive tests to be used. The pooled data gathered through the study were analyzed statistically using multilevel modeling, and the final results were adjusted for a number of factors, including children's long-standing illness, parental support for schoolwork, and home ownership. The authors noted the study's limitations: it was cross-sectional, not longitudinal; restricted to 9- and 10-year-olds; and did not focus on noise exposure in the students' homes. Moreover, the noise assessment techniques differed from country to country.

This study found that chronic exposure to aircraft noise "was associated with a significant impairment in reading comprehension. . . . [A] 5-decibel difference in aircraft noise was equivalent to a 2-month reading delay in the United Kingdom and a 1-month delay in the Netherlands" (Stansfeld et al., 2005, p. 1946). This outcome was consistent with findings from other studies on the effects of aircraft noise on reading comprehension. Because it was a cross-sectional study, the effect of long-term noise exposure to aircraft noise could not be measured. Socioeconomic status was not found to be a factor in the size of the effect. The study also found that aircraft noise was "not associated with impairment in working memory, prospective memory, or sustained attention" (Stansfeld et al., 2005, p. 1946).

The authors also looked at the effect of traffic noise on the children. In contrast to the impact of airport noise on reading comprehension, the authors noted "no effects of road traffic noise on reading comprehension, recognition, working memory, prospective memory, and sustained attention (Stansfeld et al., 2005, p. 1947).

More Sensitive Groups

Although excessive noise and reverberation in classrooms was a problem for all children, there were several especially sensitive groups for whom the problems were more acute. These groups included hearing-impaired listeners, second language learners, and children with learning difficulties.

It seems obvious that children with a hearing impairment will experience greater difficulties in conditions with excessive noise and reverberation and this was supported by a number of studies (Finitzo-Hieber and Tillman, 1978; Nábělek and Pickett, 1974). In addition, it has been reported (Reichman and Healy, 1983) that a very high proportion of students with attention and/or learning difficulties had significant hearing loss. Clearly, hearing impairment is an additional impediment to learning for children in typical classroom environments. Where a hearing-impaired child is present in a classroom, the rationale is even stronger for controlling noise levels and reverberation. Such a child can also be helped by the use of an amplification system, whereby the teacher uses an FM radio microphone that transmits directly to the listening child's hearing aid. This approach is successful because it bypasses the noise and reverberation problems of the classroom.

There are also many children who have varying degrees of fluctuating or temporary hearing impairment. These conditions are frequently due to medical problems such as ear infections or the common cold. Although temporary, while these conditions prevail, younger children, who are already at a disadvantage, will find it even more difficult to understand speech in environments with unwanted noise or reverberation.

Students for whom English is a second language are another more sensitive group that will be more strongly affected by the negative effects of noise and reverberation on their speech recognition capabilities. Nábělek and Donahue (1984) found that reverberant conditions (reverberation times of 0.8 and 1.2 seconds) reduced speech recognition scores by 10 percent in listeners for whom English was a second language. However, their use of a monaural headphone presentation technique may have exaggerated the effect. They found that second language listeners who had learned English very early in life were better able to recognize speech in noisy environments than those who had learned it later in life. Those learning English later in life need a 5 dB higher signal-to-noise ratio to perform as well as the early learners of English.

Children with learning difficulties have also been found to experience greater difficulties in understanding speech in noisy environments. Elliott et al. (1979) demonstrated that children with learning problems require louder speech (i.e., higher signal-to-noise ratios) to achieve speech recognition scores similar to those of normally progressing children. They also concluded that this effect was not due to poorer attention or hearing loss in these children. Bradlow et al. (2003) similarly found that children with learning disabilities were more adversely affected by decreased signal-to-noise ratios than were children in a control group without learning disabilities.

In classrooms having a child from one of these more sensitive groups, the rationale is even stronger for reduced noise levels and strict control of classroom reverberation to ensure that all the children can understand the words of the teacher and of their fellow students.

EXCESSIVE NOISE AND TEACHERS' HEALTH

Teachers who work in noisy classrooms must constantly raise their voices to be heard over other sounds. Over time, speaking in noisy environments can lead to vocal fatigue and other voice problems. More serious effects include the growth of nodes and polyps on the vocal cords: Often, these can be treated by voice therapy, but in some cases they require surgery (Williams and Carding, 2005). A 1993 study found that four out of five teachers who participated in the study reported some problems with vocal fatigue (Gotaas and Starr, 1993). A 1995 study of populations in the U.S. workforce that rely on voice as a primary tool of their trade found that teachers constitute more than 20 percent of the voice-clinic load, or five times the number expected by their prevalence in this segment of the workforce (that is, the segment that relies heavily on the use of their voices) (Titze et al., 1996). Similar results were found by a Swedish study (Fritzell, 1996).

Work by Preciado et al. (1998) is one of the few efforts to attempt to relate vocal disorders in teachers to environmental factors. Voice problems occurred more frequently for teachers of the lowest grades, those in larger classrooms, those with more students, and those in classrooms with higher noise levels. Smith et al. (1998) found that while 20 percent of the teachers had missed work owing to voice problems, only 4 percent of other professionals (nonteachers) had done so. They said their findings suggest that "teachers are at a high risk of disability from voice disorders and that this health problem may have significant work-related and economic effects" (p. 480).

SOLUTIONS/DESIGN REQUIREMENTS

There is clear evidence that noise and reverberation in typical classrooms interfere with children's ability to understand speech. These problems are not well recognized by those unfamiliar with the many research studies on the topic. They are also not well recognized because adults do not suffer the same communication problems in classrooms as do young children. Further, it is reasonable to suppose that there are connections between impaired communication conditions in classrooms and the many reports of decreased student achievement in noisy situations. These problems can be mitigated by school designs that employ quiet mechanical systems (heating, ventilating, plumbing, etc.) and isolation against noise from adjacent spaces and outdoors. Effective and supportive classroom acoustics can be achieved by paying attention to the construction of a classroom, the sources—external and internal—of noise, and the children's language and learning needs.

When planning schools, criteria for ambient noise level should be set, designed for, and eventually verified through a commissioning process. Noise criteria can be based on A-weighted sound levels or on several other common measures of noise such as values of noise criterion (NC), room criterion (RC), or balanced noise criterion (BNC). In addition, the desired room reverberation time should be established to achieve room acoustical quality appropriate for speech communication.

Careful HVAC design and installation are needed to meet requirements for background noise from HVAC sources, especially where sound-absorbing duct liners are not being used. Designers should consider using quiet fans, fan silencers, duct vibration "breaks," and duct systems that deliver air to classrooms at sufficiently low velocities to minimize flow noise. In general, unit ventilators are not nearly quiet enough to meet the ambient noise criteria.

Achieving adequate sound isolation of a classroom against noise transmitted from adjacent spaces is also important. Designing to reduce unwanted sounds from adjacent spaces is done by specifying appropriate sound transmission class (STC) values for the partitions separating a classroom from adjacent spaces. Reducing noise transmission from outdoor sources is equally important and should be achieved through appropriate site selection (e.g., away from persistent transportation noise). In addition, HVAC equipment should be selected to meet acceptable outdoor noise levels and should be located away from classrooms. Windows will be a critical component in controlling the intrusion of outdoor noises.

Specifying the acoustical design criteria requires some special effort. The criteria can most easily be determined by following the recommendations of ANSI Standard 12.60, "Acoustical Performance Criteria, Design Requirements, and Guidelines for Schools." Tables 6.1 and 6.2 summarize some of the recommendations of this standard.

TABLE 6.1 Required Acoustic Conditions in Classrooms

Condition	Recommended Allowable Maximum
Ambient noise level	35 dBA
Reverberation time	0.6 s

SOURCE: ANSI S12.60.

TABLE 6.2 Sound Isolation Requirements Between a Classroom and Various Types of Adjacent Spaces

Type of Adjacent Space	STC Requirement for Separating Partitions
Another classroom	50
Washroom	53
Corridor/conference room	45
Music/mechanical room	60
Outdoors	50[a]

[a]In the case of outdoor noise, a site-specific design is recommended.
SOURCE: ANSI S12.60.

In addition to the requirements for the various physical components of the classroom and other parts of the school building, it is also very important that teachers appreciate the effects of noise and poor acoustics, how these effects vary with the age of children, and how they can manage their teaching activities to best communicate with the students.

CURRENT GREEN SCHOOL GUIDELINES

Current green school guidelines typically indicate some recognition of the importance of outdoor and indoor noise management in classrooms. They recognize the measurable impact of noise on the academic performance of students and its interference with speech communication in classrooms.

In some cases, green school guidelines call for the reduction of background sound levels to NC35 or less, to NC30. However, only NC30 approximately meets the requirements of ANSI S12.60, which is intended to ensure that a teacher's voice is clearly understood by younger children against a background of other local noise-generating equipment or activities.

Some green school guidelines recommend meeting ANSI S12.60 standards for sound isolation between classrooms and adjacent spaces so that noise from these spaces does not compromise the quiet background sound levels needed for a teacher to be clearly understood.

Current green school guidelines do not fully address controls on outside noise generation. Given the results of research on the impact of traffic and airplane noise on student performance, these sources and the selection of school sites to avoid loud outdoor sound need to be considered.

FINDINGS AND RECOMMENDATIONS

Finding 6a: Most learning activities in school classrooms involve speaking and listening as the primary communication modes. The intelligibility of speech in classrooms is related to the levels of speech sounds relative to the levels of ambient noise and to the amount of reverberation in a room.

Finding 6b: Sufficient scientific evidence exists to conclude that there is an inverse association between excessive noise levels in schools and student learning.

Finding 6c: The impacts of excessive noise vary according to the age of students, because the ability to focus on speech sounds is a developmental skill that does not mature until about the ages of 13 to 15 years. Thus, younger children require quieter and less reverberant conditions than do adults to hear equally well. As adults, teachers may not appreciate the additional problems that excessive noise creates for younger students.

Finding 6d: Excessive noise is typically a more significant problem than is too much reverberation in a classroom. It is not possible to have both increased speech level (to maximize signal to noise) and reduced reverberation times. Good acoustical design must be a compromise that strives to increase speech levels without introducing excessive reverberation.

Finding 6e: The most substantial body of research related to excessive noise and learning in the classroom addresses the impacts of road traffic, trains, and airport noise.

Finding 6f: Some available evidence indicates that teachers may be subject to voice impairment as a result of prolonged talking in noisy school environments. However, there is no information to quantify a relationship between specific noise levels in classrooms and potential voice impairment.

Recommendation 6a: To facilitate student learning, future green school guidelines should require that new schools be located away from areas of higher outdoor noise such as that from aircraft, trains, and road traffic.

Recommendation 6b: Future green school guidelines should specify acceptable acoustical conditions for classrooms and should require the appropriate design of HVAC systems, the design of walls and doors separating classrooms and corridors, and the acoustic quality of windows and walls adjoining the outdoors. This recommendation is most easily achieved by requiring that green schools comply with American National Standards Institute (ANSI) Standard 12.60, "Acoustical Performance Criteria, Design Requirements, and Guidelines for Schools."

Recommendation 6c: Additional research should be conducted to define optimum classroom reverberation times more precisely for children of various ages.

7

Building Characteristics and the Spread of Infectious Diseases

Transmission of infectious diseases among school children is frequent and probably inevitable. Although most such cross-infections are transient and not serious, they may have concentric circles of repercussions for other students, teachers, parents, and educational achievement. Actions that can help control the spread of such infections are, therefore, important to the health of students, teachers, and other adults.

Outbreaks of acute infections caused by viruses, bacteria, parasites, and protozoa have been documented among both students and staff in schools (Harley et al., 2001; Brennan et al., 2000; Khetsuriani et al., 2001; Davison et al., 2004; Lee et al., 2004; Otsu, 1998; AAP, 2003a,b). The majority of infections causing absenteeism in schools are respiratory and gastrointestinal illnesses. These types of illnesses occur continuously throughout the year. Although they are less obvious than outbreaks of infections, they nonetheless have an impact (Hammond et al., 2000; Brady, 2005; AAP, 2003a,b). Respiratory infections acquired in schools, especially rhinovirus infections, play a major role in exacerbating asthma and causing hospitalization for it (Johnston et al., 1995, 1996). Such illnesses result in increased health care costs, lost public funding for schools, added administrative expenses, and absences from work on the part of parents.

For children attending public schools, the transmission of common infectious diseases is estimated to result in more than 164 million lost school days each year (Vital Health and Statistics, 1998). On average, students are absent from school 4.5 days per year, and teachers miss work 5.3 days each year because of acute illness (Middleton, 1993). When surveyed, more than 80 percent of teachers cited school absenteeism as their main problem (Carnegie Foundation, 1990). This finding is complemented by research showing the detrimental effect of absenteeism due to illness on student achievement (Ohlund and Ericsson, 1994).

Infectious agents can be transmitted by person-to-person contact, by droplets or large particles spread by coughing, sneezing, or wheezing, or by fomites (self-innoculation after touching contaminated surfaces). Distinguishing the various transmission modes is difficult, and the role of airborne transmission in respiratory infections is not yet completely understood. It is clear that many factors affect infection transmission, including the number of students in a school, their ages, the degree of crowding in classrooms, teacher-to-student ratios, geographic location, seasonal variation in the activity of common infectious agents, the number of recent immigrants, the proportion of children with chronic health conditions, and the number of young children: Young children have relatively less immunity to common infections and tend to pay less attention to personal hygiene (Hall, 2001; Marchisio et al., 2001; Goldmann, 2001; Brady, 2005; Musher, 2003).

School building factors that influence the transmission of specific infectious agents include the ventilation rate, humidity, filtration effectiveness, the cleanliness of surfaces, and the number, accessibility, and functional state of sinks and toilets (Bloomfield, 2001; Vernon et al., 2003) (Figure 7.1).

Current green school guidelines typically contain few measures related to these building characteristics, although they may address filter efficiencies. Additional measures could be included in future guidelines that might interrupt the transmission of infectious diseases, thereby improving the health of students, teachers, and others.

MODES OF TRANSMISSION FOR RESPIRATORY VIRUSES

The main agents of respiratory infections in schoolchildren, their families, and caretakers are respiratory syncytial virus,[1] picornaviruses (especially the three genera enteroviruses, rhinoviruses,[2] and parechoviruses[3]), influenza, adenoviruses,[4] parainfluenza viruses,[5] measles, and

[1]A virus that causes infection of the lungs and breathing passages. Symptoms can include cough, stuffy or runny nose, mild sore throat, fever.

[2]The common cold.

[3]Causes mainly gastrointestinal and respiratory infections.

[4]A group of viruses that infect the membranes (tissue linings) of the respiratory tract, the eyes, the intestines, and the urinary tract.

[5]Parainfluenza viruses account for a large percentage of pediatric respiratory infections, including upper respiratory tract infections, laryngotracheobronchitis (croup), bronchiolitis, and pneumonia.

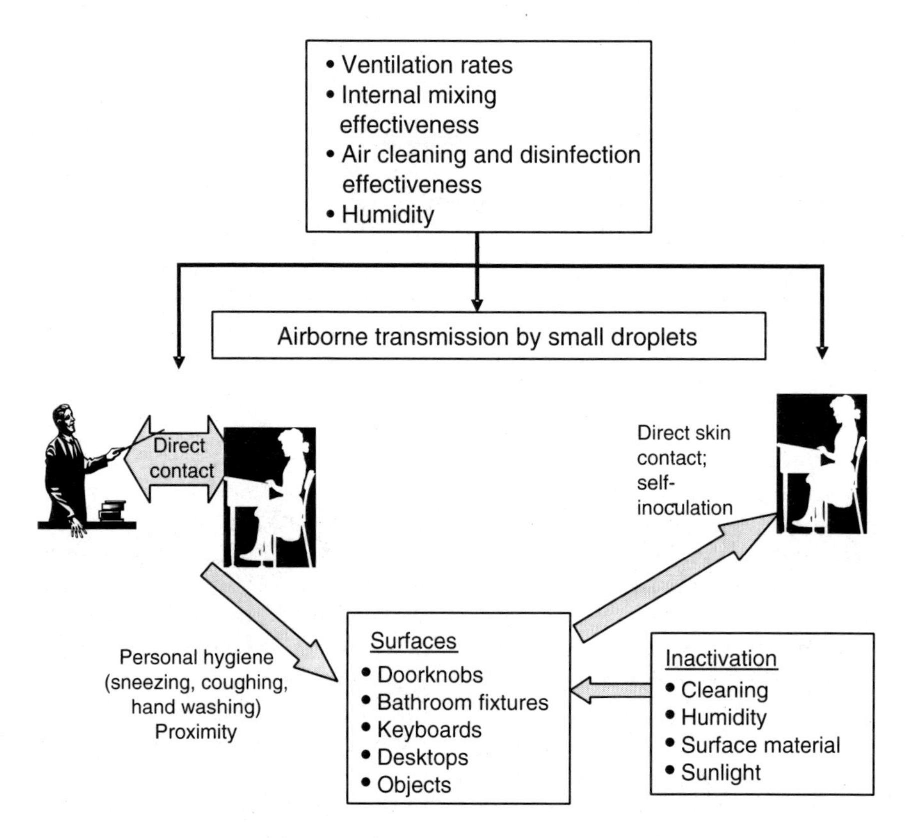

FIGURE 7.1 Transmission modes for spread of infectious diseases.

human respiratory coronaviruses[6] (McIntosh, 2005). Respiratory viruses are shed primarily as nasal or oral secretions (runny nose) or projected into the air by sneezing, coughing, spitting, or speaking. Most are transmitted from the infected person to others by any of three possible mechanisms mentioned above. Specific viruses may be spread by more than one of these mechanisms.

Small-particle aerosols (<10 μm mass median diameter) are usually generated by coughing or sneezing and may traverse distances ≥1.8 m, can occur without close or direct contact with the infected person or the

[6]Human coronaviruses (HCoV) cause common colds but can also infect neural cell cultures.

infectious secretions. Evidence is emerging that quiet breathing can also generate small particles. Aerosolized respiratory fluids (small droplets, sometimes less than 1 μm in size) evaporate quickly when discharged into room air, resulting in droplet nuclei (Riley and O'Grady, 1961). Such small particles do not settle on surfaces and clothes to any appreciable extent, are inefficiently removed by conventional filtration, and are easily dispersed by air movement indoors. Viruses that can be spread by small-particle aerosols, such as the measles, chicken pox, and, sometimes, influenza, can cause rapid outbreaks of infection in susceptible populations.

Transmission by large droplets or large particles, in contrast, requires close person-to-person contact, usually at a distance of ≤0.9 m, for infection to occur directly: Large droplets (>10 to 100 μm diameter) expectorated during a sneeze or cough may be released with sufficient momentum to contact directly the mucosa of another person's eyes, nose, or mouth. However, large droplets have appreciable settling velocities and infectious material might also land on clothing or the surfaces of nearby objects. Some viruses are able to remain infectious on environmental surfaces, are transferable to skin, and can remain infectious for a long enough time to allow self-inoculation into the respiratory tract (Hall, 1981, 2000).

The third transmission mechanism is self-inoculation by touching contaminated surfaces. Clinical studies, especially those using infection control interventions, have indicated that the primary mode of transmission for the most common respiratory and gastrointestinal viruses acquired by children and adults within schools is close contact with infected individuals or touching infectious secretions that contaminate inanimate objects (Hall, 1981; Hall and Douglas, 1981; Hall et al., 1981; Ansari et al., 1991; Mbithi et al., 1992). These articles suggest that fomites are a primary mode and that large droplet spread (across distances of less than 3 feet) is also important.

Fabian et al. (2006) point out that discrimination between the various modes of transmission has been difficult for several reasons. One is that the concentration of infectious particles in air required to transmit infection is very low—thus it can be very difficult to detect them and prove that the particles exist. A second reason is the nature of human source aerosols and their dissemination has not been well understood or widely appreciated.

Understanding the role of airborne transmission is further complicated by the fact that the human infectious dose will vary substantially among viruses. Dilution in the indoor air supply may make concentrations far too low for transmission to occur. However, there is no reason to believe that airborne viruses are uniformly dispersed in indoor spaces. The mixing of air within a classroom and certainly within spaces having cubicles, partitions, files, and bookshelves is far from homogeneous. Areas of high concentrations might persist for longer than would be predicted by the calculated air exchange rate for the room. Reports on the transmission of respiratory infections, including SARS, in places like airplanes, office buildings, and hospitals also suggest the potential importance of an airborne component (Moser et al., 1979; Roy and Milton, 2004; Reichert et al., 2001; McLean, 1961; Riley et al., 1978).

New insight into the generation of respiratory aerosols and the role of airborne transmission in respiratory infections is being gained with more sophisticated aerosol measurement equipment (Fabian et al., 2006). Earlier work focused on larger droplets produced from the upper respiratory mucosa (Riley and O'Grady, 1961) and used techniques generally insensitive to particles less than 1.0 μm in diameter. Those early studies found that as many as 40,000 droplets are generated during a sneeze, about one order of magnitude more than by coughing and two to four orders more than by talking loudly (Cox, 1987). Most of the particles detected were less than 4 μm (Duguid, 1946; Gerone et al., 1966). Very few particles were thought to be liberated during quiet breathing (Riley and O'Grady, 1961), and there were few data to suggest interindividual differences in the generation of aerosols during coughing, sneezing, or talking.

More recent studies conducted by Edwards et al. (2004) show it is possible to generate large numbers of particles from the human respiratory system during quiet mouth breathing. They used a laser light-scattering instrument (not available to earlier researchers) to detect particles in the submicron range down to 150 nm. Although the number of subjects was small (11), they demonstrated a clear bimodal phenomenon with respect to respiratory particle generation during quiet breathing. Over the 6 hours of monitoring, five volunteers produced an average of only 14 to 71 particles per liter of exhaled air. However, the other six volunteers consistently exhaled 10 times that number of particles per liter (an average of more than 500 particles per liter over 6 hours), with one subject generating up to 3,230 particles per liter during quiet breathing. This translates to 5,000 to 32,000 particles liberated per minute for those "high producers." The mechanism by which these submicron droplet nuclei are generated is not known. Microdroplets formed by bursting films of respiratory fluid or the fragmentation of specula during breathing are suspected of producing submicron droplets. Obviously these intriguing observations need further research to determine if they might explain the existence of "super shedders" in outbreaks of SARS, measles, and influenza.

Roy and Milton (2004) characterized the potential aerosolized transmission as "obligate" (TB), "preferential" (measles), or "opportunistic" (most respiratory viruses). This classification is based on an agent's capacity to be transmitted and to induce disease through fine-particle aerosols and other routes. The authors noted that probably all viruses have different likelihoods of being spread by aerosols or droplets but the likelihood of spreading is highly variable and the concentration of virus is often too low to cause human infection, despite its being detectable by advanced techniques (Roy and Milton, 2004). A number of clinical studies suggest that the primary or "preferential" mode of spread is not airborne. For example, it seems that rhinoviruses can be airborne, although the preponderance of real-world studies suggested that this was not the preferential mode of spread (Hendley et al., 1973; Winther et al., 1998; Gwaltney and Hendley, 1982; Gwaltney et al., 1978). A well-designed experiment by Jennings and Dick (1987) provided evidence that the airborne route was more important for the transmission of rhinovirus than contact and fomites. Myatt et al. (2004) reexamined earlier studies of the Gwaltney group and found them to be underpowered to detect airborne transmission.

The importance of this evidence to schools administrators, teachers, and other decision makers is that they need to recognize multiple routes for the transmission of infectious diseases. As discussed in the following sections, control strategies range from the mundane and inexpensive—control via hand cleansing and decontamination of surfaces—to improved ventilation systems, including displacement ventilation.

MEASURES FOR CONTROLLING THE SPREAD OF INFECTIONS IN SCHOOLS

Two of the most effective measures for controlling the spread of infection in schools are immunization and the exclusion of children and adults with infectious cases. Because these are policy measures unrelated to school buildings, they are not discussed further in this report.

Design-related control measures have generally focused on hand cleansing, along with personal hygiene and decontamination of surfaces. Most studies evaluating methods of infection control have been in recognized high-risk populations, such as hospitalized and compromised patients and children in out-of-home care who are less than 3 years old. Whether the findings of these studies can be applied to the general population of K-12 students and staff in schools is unclear. The exception is the benefit of effective hand cleansing. Some studies suggest that effective and routine hand washing can diminish common infections and could cut down on absenteeism among school children (Kimel, 1996; Master et al., 1997; Early et al., 1998). Evident also, however, is the difficulty of maintaining compliance in a school population (Guinan et al., 1997, 2002), especially among younger children.

Hand and Personal Hygiene

The use of hand sanitizers or hand rubs for infection control been evaluated in health care facilities, at home, and in day care centers and schools (Meadows and Le Saux, 2004; Lee et al., 2005; Girou et al., 2002; CDC, 2002; Rochon-Edouard et al., 2004; Master et al., 1997; Early et al., 1998; Zaragoza et al., 1999; Turner et al., 2004; Sandora et al., 2005). Most of these studies showed a significant reduction in the occurrence of infections in general. However, effectiveness of the sanitizers may vary depending on the individual agents, especially viruses (IFH, 2002b). Hand sanitizer products containing alcohol tend to have greater activity against nonenveloped viruses (e.g., respiratory syncytial virus, influenza, parainfluenza viruses, HIV, and herpes simplex virus) than against enveloped viruses (e.g., rhinovirus, parechoviruses, enterovirus, adenovirus, rotavirus,[7] and caliciviruses).[8] Washing one's hands for an extended time, however, can enhance the antiviral activity of hand sanitizers toward enveloped viruses. Picornaviruses (e.g., enteroviruses, rhinoviruses, and hepatitis A virus) are particularly difficult to eradicate from hands or environmental surfaces but may be inactivated by high concentrations of alcohol and extended hand-washing time. Nevertheless, for all of these viruses, hand sanitizers have been shown to be more effective than routine hand washing (IFH, 2002a,c).

A number of studies have examined the efficacy of hand sanitizers over traditional hand-washing methods in reducing the incidence of acute illnesses and absenteeism. A systematic review of five published studies on the effectiveness of hand sanitizers in preventing absenteeism due to illness in elementary school children revealed a significant advantage for hand sanitizer use over baseline conditions, which may have included soap and water or no structured hand washing. However, the quality and heterogeneity of the reports prevented the calculation of pooled estimates (Meadows and Le Saux, 2004).

Recently the Centers for Disease Control and Prevention (CDC) revised its guidelines for hand hygiene in health care settings to recommend hand washing to remove visible soiling. Otherwise the use of hand sanitizers is recommended based on their efficacy, the propensity of health care workers to use them, and the low incidence of adverse effects (CDC, 2002).[9]

The CDC's recommendations also emphasized that for health care facilities the items necessary for hygiene and cough etiquette must be

---

[7]Rotavirus is the most common cause of severe diarrhea among children.

[8]Caliciviruses produce infections that cause acute diarrhea and vomiting (gastroenteritis), abdominal cramps, myalgias, malaise, headache, nausea, and low-grade fever.

[9]Adverse effects of hand washing can include skin irritation and dryness, especially when these conditions are pre-existing.

readily available, including disposable tissues, no-touch waste receptacles, conveniently located dispensers of alcohol-based hand rub solutions, and, wherever sinks are available, supplies for hand washing. Similar measures could be implemented in green schools; ongoing maintenance would be required if such measures are to be effective.

The efficacy of the measures contained in these recommendations in reducing respiratory illness and absenteeism has not been proved. The recommendations are based, instead, on the assumption that the measures will disrupt the primary modes of transmission of the agents most frequently causing respiratory and gastrointestinal illnesses. Assuring compliance with these measures is problematic for any age and probably impossible for young children. However, for older children and staff, constant reminders to follow these measures and ensuring that hand sanitizers and other items are routinely available seem reasonable considering the ways in which infectious agents spread, the low cost of these items, and the absence of significant adverse effects.

DECONTAMINATION OF ENVIRONMENTAL SURFACES

Many of the common agents causing infections in schools have been found to survive on hands and environmental surfaces long enough to allow the transfer of infectious organisms from the environment to the hands. The duration of viability and the inoculating dose required are highly variable according to the organism. A number of studies, mostly under experimental conditions, have evaluated the survival of common organisms on inanimate objects in order to ascertain their potential for being spread by fomites (Hall et al., 1980; Brady et al., 1990; Butz et al., 1993; Dettenkofer et al., 2004; Dowell et al., 2004; Neely and Orloff, 2001; Rogers et al., 2000; IFH, 2002b). Few of these studies, however, went so far as to determine the survival of the transferred organisms on the skin or under different environmental conditions (Hall et al., 1980).

Surface decontamination procedures include cleaning using a liquid or soap detergent and hot water. In general this method has been found to produce a surface clean enough that the organisms that remain are not transferable or are too few in number to cause human illness. Rinsing well after cleaning is a critical step. This method of cleaning may be less effective if the surface is damaged or in poor condition. In addition, for this or any other method, results will be compromised by using a drying cloth or other material that is not initially clean and that is used continually without itself being repeatedly decontaminated.

Surfaces are also cleaned using a soap or detergent followed by an application of a chemical disinfectant product. The disinfecting product should achieve a hygienic surface sufficient to interrupt the spread of organisms to cause infection. Furthermore, the agent should have a broad and rapid inactivating action against a variety of bacterial, viral, and fungal pathogens.

Use of a combined detergent and disinfectant product allows for a one-step procedure. The characteristics of disinfectants in the combined products, such as their recommended concentrations, whether they are static or cidal toward the microbes, and the recommended methods of use, are published or available on the Web, along with guidelines for the selection and use of disinfectant products (IFH, 2002a; Rutala et al., 2000; Rutala and Weber, 2004).

Literature comparing the effect on cross-infection rates of disinfecting environmental surfaces as opposed to cleaning them without using disinfectants was systematically reviewed (Dettenkofer et al., 2004). Of the 236 identified published articles, none was a randomized controlled trial and only four were cohort studies with matching inclusion criteria. None of these studies showed diminished rates of infection associated with the routine practice of disinfecting surfaces in comparison to cleaning them with detergent only. The authors concluded that disinfectants could adversely affect individuals in the area and also the environment and that if they were used, special safety precautions should be taken. The authors also concluded that more well-designed studies would need to be performed to assess the role of surface disinfection in preventing the transmission of infection.

One alternative to conventional methods of disinfection of environmental surfaces would be the development of hands-free designs for those surfaces most often touched—doors, faucets, flushers, soap and towel dispensers, and entrances to restrooms. While these solutions are becoming more frequent in airports for their ease of cleaning, they are less common in schools. Research on the effect of hands-free design on infection rates in schools would be valuable.

Choosing surfaces and components such as faucets, door handles, and so forth, that can be easily cleaned may help to interrupt the transmission of infectious diseases through fomites. One precedent of note is the involvement of "cleaning engineers" (a discipline in Holland) in the review of design/engineering drawings. Their signature certifies that the building as designed can be effectively cleaned and will not create undue health and degradation risks.

VENTILATION AND AIR CLEANING

Three building characteristics are relevant to the survival, dispersal, and removal of airborne pathogens—the relative humidity of the room air, the dispersion path and dilution provided by the ventilation system, and the return air mixing and filtration provided in the fresh air supply.

Exhaled particles will readily evaporate in typical indoor humidities, and disperse or slowly settle on surfaces and people. Aerosols even 5 μm in size have slow settling velocities, taking about 44 minutes to settle from a height of 2 meters. A 1 μm particle would take 18 hours to settle in still air. Since settling is negligible for particles less than 5 μm in size, they will disperse with the air currents created by mechanical ventilation systems, movement of people, buoyant convections by people and equipment, and pressure-driven flows (natural and mechanical) within a room, throughout a structure, and even into the outdoor air.

Humidifiers are not routinely installed in schools. Although humidification carries its own health concerns, a lack of it in school buildings can result in low relative humidity (RH) (<30 percent) in the winter (heating months), although the RH would be more moderate (30-70 percent) during periods when the outside air must be cooled. At lower relative humidities common to schools, exhaled drops would quickly evaporate and be suspended for long periods before finally settling onto surfaces or being carried out of the room or building by the exhaust air. Furthermore, low humidity has been associated with increased viability for some viruses. Harper (1961) showed that vaccinia and influenza viruses survived longer at RHs in the low 20 percent range. Other viruses have optimal survival at different RH values. Remington et al. (1985), in a study that followed up on a measles outbreak in a doctor's office, suggested that low humidity was a contributing factor. Given low RHs and hard surfaces, viruses might survive as suspended droplets or on fomites for many hours in schools. On the other hand, the rhinovirus (common cold) survives longer at moderate to high relative humidities.

A second factor affecting the spread of infections in schools is the dispersion path and dilution provided by the ventilation system. The conventional design for HVAC systems is to forcefully introduce conditioned air into a room at the ceiling level to ensure effective mixing of "fresh" air with the existing room air. The mixing within a room (or zone), however, will in turn cause the dispersion of indoor-generated viral material present as small droplets. In office environments, Myatt et al. (2004) collected nasal swabs and air samples to look for viruses in nearby colleagues using the polymerase chain reaction. They confirmed that an office worker was responsible for a virus strain isolated in an air sample taken in the same general work area.

Although conventional ceiling-based ventilation systems can cause dispersion of viral material, increasing ventilation rates can speed the dilution and removal of viral material. Milton et al. (2000) tracked sick leave as a function of CO_2 levels and building ventilation. Across 42 buildings grouped according to ventilation rates, employees in buildings with higher ventilation rates (approximately 48 cfm/person diluting outside air) reported using 1.6 fewer days of sick leave than employees working in the buildings at lower ventilation rates (4 cfm/person the current ASHRAE 62.1 2004 standard).

While increasing the ventilation rate may cause more rapid mixing, it also causes more rapid dilution and removal. However, there may be practical limitations to this effect, since not all the air is completely replaced with each air exchange. Typically only two-thirds of the air is removed per air exchange because of the size and location of supply and exhaust ducts, short circuiting, and obstructions (blocked unit ventilators). Also, dilution with increased ventilation air can cost more for larger ducts, fans, and heating and cooling. The potential for increased mechanical noise in classrooms would also be a concern (see Chapter 6).

An alternative to this typical mechanical mixing of conditioned air from the ceiling is the use of floor-based or side-wall-based displacement ventilation. Theoretically, ventilation effectiveness is maximized by displacement flow. By bringing in fresh air near the floor and returning it at the ceiling, infectious droplets and contaminants are not forced toward neighboring occupants. As demonstrated in chamber studies, exposures to occupants can be reduced substantially and ventilation efficiency can be increased.

A third ventilation-related factor affecting the spread of infections within schools is the percent of the return air mixed into the supply air and the filtration provided in the air stream. Most schools are conditioned with a combination of recirculated air and outside air that has either been cooled, with some ambient moisture removed, and/or heated to meet thermal requirements.

To reduce the potential for exchanging infectious substances, schools could be designed to use 100 percent outside air, as hospitals have done. To ensure energy efficiency, however, it would be critical to introduce heat recovery or to separate the ventilation from the thermal conditioning system. Thoughtful attention to thermal heat loads, sensible and latent heat recovery, and possibly the use of thermal elements such as chilled beams and radiant heating might make affordable the use of 100 percent outside air to meet ventilation requirements, exhausting all breathing air that might contain infectious aerosols. Research on the costs and benefits of eliminating recirculated air in schools should be undertaken for different climate conditions.

At the same time, it is imperative to assess the effects of filtration on reducing airborne pathogens and pollutants from both outside and return air sources. Airborne pathogens that have evaporated to submicron size will not settle out nor will they be effectively removed by filtration.

Indeed, in a recirculated air system, viable airborne pathogens could be distributed to other classrooms by the mechanical ventilation system. Riley et al. (1978) present the case of a measles outbreak in an elementary school consistent with airborne transmission. Riley and Nardell, in their 1989 review, cite cases where tuberculosis was transmitted through the air to infect people who had no direct contact with the infected person.

Germicidal UV Radiation

Viruses can be readily inactivated with ultraviolet (UV) light (Nuanualsuwan et al., 2002; Nuanualsuwan and Cliver, 2003), and the use of UV light to control the airborne transmission of infectious respiratory agents has been documented (Riley and O'Grady, 1961; Riley and Nardell, 1989). Germicidal UV radiation in the short wavelengths (100 to 280 nm) is referred to as UVC radiation or UVGI. The Centers for Disease Control and Prevention's (CDC) Web site (www.cdc.gov) cites numerous instances of UVC use to kill or inactivate tuberculosis and other infectious agents in hospitals, military housing, and classrooms. UVC fixtures have been used to clean the air in areas where undiagnosed tuberculosis patients might be present, such as emergency room waiting areas, shelters, and correctional facilities.

More than 40 years ago, Jensen (1963) demonstrated the germicidal effectiveness of UV radiation at 254 nm wave length (UVC) on vaccinia and other viruses. The work of McClean (1961) is often cited as evidence that UV irradiation has germicidal efficacy for airborne viruses. During the late 1950s, a Veterans Administration hospital in Livermore, California, was operating upper room UV lights in one of its wings. Two outbreaks of acute febrile respiratory illness (ARI) occurred in two wings of the hospital during the fall and winter of 1957/58, one of them in the wing with the UV lights. Baseline serology was obtained in July 1957 before the first outbreak. Follow-up serology was taken in November 1957 and again in March 1958 after the first and second ARI outbreaks. Only staff or visitors could have been the source of infection since the patients were in long-term care and restricted to the hospital. Since the source of the first infection was not known, it was also not known whether patients in both wings were similarly exposed. Nevertheless, there were four times as many influenza titers[10] in the wing without the upper room UV lights. Over the two outbreaks, 18 percent of the staff and 19 percent of the patients in the wing without the UV lights became ill, whereas only 2 percent in the wing with the upper room UV lights were ill. McClean reports that the

[10]A titer is the unit in which the analytical detection of many substances is expressed. It is the result of a titration.

first wave of ARI was not influenza but that influenza clearly dominated (not exclusively) the second outbreak. Nevertheless, the UV lights in one of the wings were protective against both viral agents. There have not been any additional epidemiologic studies on infection control with UVC in hospitals, offices, or schools.

Today's increased focus on technologies that offer protection against biological threats has renewed interest in the germicidal effectiveness of UVC. This, in turn, has led to recent chamber studies in which UV lamp placement or configuration, room air exchange, supplemental air mixing, wattage (UV fluence rates in μmW/scm^2), and relative humidity have been looked at to determine the rates at which they inactivate several types of spores, cells, and viruses. The CDC Web site (www.cdc.gov) summarizes the findings in its 2005 TB guidelines published in *Morbidity and Mortality Weekly Report*, December 30, 2005. The salient facts are described below.

In a well-mixed room, UVGI fixtures properly located to irradiate the upper space of a room (>8 feet) and having the appropriate wattage can significantly enhance the inactivation rate of test organisms. A substantially higher "effective" air exchange rate (ACH) can be achieved with a combination of upper room UV and mechanical air change rates. Effective ACHs with upper room UVC can greatly exceed what is practically possible with other combinations of air filtration and mechanical ventilation (Nardell et al., 1991). However, inactivation effectiveness diminishes as UV fluence decreases and as RH increases above 60 percent. Servicing of UV lamps to maintain performance specifications is important, particularly in settings when the lamp surfaces can become soiled. Guidelines for application of upper room UVC are contained in several ASHRAE publications (First et al., 1999a,b).

The long-term exposure of occupants to UVC radiation is a concern because it could cause skin and eye damage. The CDC and the National Institute for Occupational Safety and Health (NIOSH) have published recommended exposure limits to protect health care workers from the acute effects of UV light exposure. The potential carcinogenic effects of UVA and UVB, but not UVC, are reviewed by Parrish (2005). The exposure of building occupants to UVC used in upper room installations is very low. A recent report on UVC dose monitoring showed that even when some areas of the lower room exceeded 0.2 μW/cm^2, the 8-hour dose was a small fraction of the 6 mJ/cm^2 recommended exposure limit (First et al., 2005). There is a risk of acute photokeratitis, a self-limiting condition. This is the basis of the 6 mJ/cm^2 8-hour maximum dose standard. There is no literature demonstrating a long-term risk to the eye from the extremely low exposures to UVC in the lower room. Even if one assumed that an installation in a school classroom gave an average exposure of 0.2 μW/cm^2 for a 6-hour school day, every day for a 180-day school year, the total UVC dose to the skin from the UVC lights would be $<0.8\ J/cm^2$. By comparison, the average adult receives $33\ J/cm^2$ of UVA and UVB from sunlight each year. Given that 50 percent of UVA and UVB penetrate to the basal cell layer while only 5 percent of UVC penetrates to expose the dividing cells, the maximum conceivable annual dose to the basal cells of the epidermis from use of upper room UVC in classrooms would be 0.24 percent of the annual dose of UVA plus UVB from sunlight.

When UVC lamps have been placed in ventilation ducts or in the air-conditioning systems of air-handling units (cooling coil and drip pans) or have been configured into portable room air cleaners, the UVC exposure to occupants has essentially been eliminated. A chamber study of portable UVGI systems demonstrated their effectiveness in inactivating or killing airborne vegetative bacteria (Green and Scarpino, 2002). The efficiency of UVGI in air-handling units was tested in an office building and reported by Menzies et al. (2003), who examined microbial contamination including endotoxins but not airborne pathogens (see Chapter 4). The application of supplemental air cleaning with UVGI for classrooms, while promising, should first be researched and evaluated more thoroughly. Although enclosing UVC in air ducts or portable air cleaners might reassure parents, its effectiveness will be limited by the amount of supply air that can be treated. Upper room UVC still has a comparative advantage because more effective air exchange rates can be achieved (Nardell et al., 1991).

FINDINGS AND RECOMMENDATIONS

Finding 7a: Common viruses and infectious diseases can be transmitted by multiple routes: through the air, by person-to-person contact, and by touching contaminated surfaces (fomites). Certain characteristics of buildings, including the cleanliness of surfaces, relative humidity, and ventilation effectiveness, influence the transmission of common viruses. Evidence from studies in nonschool environments suggests that interventions which interrupt the known modes of transmission of common infectious agents may decrease the occurrence of such illnesses in schoolchildren and staff.

Finding 7b: The best way to control infections, especially gastroenteritis, appears to be instituting procedures that promote good hand cleansing. Available but limited information indicates that hand sanitizers are superior to the routine washing of hands.

Finding 7c: Cleaning of surfaces that are commonly touched (e.g., doors, faucets, desktops) is effective for interrupting the transmission of infectious agents. Disinfecting surfaces with water and detergents is apparently as effective as applying germicidal agents.

Finding 7d: The use of no-touch faucets, doorways, receptacles, and equipment seems to be a reasonable, though unproven, method for infection control.

Finding 7e: The survival, dispersal, and removal of airborne pathogens are affected by relative humidity, ventilation rate, and the percentage of recirculated air in the air supply. Increased ventilation rates have been shown to speed the dilution and removal of viral material. The use of displacement ventilation and the reduction of the percentage of recirculated air in the air supply have the potential to reduce building occupants' exposures to airborne pathogens.

Finding 7f: Ultraviolet germicidal irradiation (UVGI) may be effective for inactivating and killing some infectious organisms, but its use in schoolroom applications has not been systematically studied.

Recommendation 7a: Future green school guidelines should include measures for the regular cleaning of commonly touched surfaces and the availability of hand sanitizers at sinks. The use of "no-touch" faucets, receptacles, equipment, and egress from bathrooms should be considered, taking into account the age of the children in the school.

Recommendation 7b: Full-scale classroom and school studies should be conducted to quantify the efficacy of a variety of ventilation strategies, including displacement ventilation and the elimination of recirculated air, for the dispersion and removal of airborne infectious agents. Studies should also quantify the potential costs and benefits of such ventilation strategies.

Recommendation 7c: Additional research should be conducted to determine the optimal infection-control interventions in terms of measurable outcomes such as absenteeism and academic achievement. One line of research is the use of ultraviolet germicidal irradiation in supplemental or portable air-cleaning devices in schoolroom applications and its effects on human health.

8

Overall Building Condition and Student Achievement

Professional organizations and governmental agencies have for many years been calling attention to the deteriorating condition of the nation's schools. The importance of effective operations and maintenance practices to the satisfactory performance of a school building's envelope, mechanical systems, and surfaces has been emphasized in previous chapters.

The committee identified 20 studies that investigated the relationships between *overall* building condition and student achievement and *overall* building functionality and student achievement. The identified studies primarily come from the field of educational research and investigate possible relationships between individuals and groups and their physical environment.

In the course of its review, the committee identified significant limitations in the methodologies used in these studies (the limitations are discussed in a section near the end of this chapter). Nonetheless, the committee believes there is value in describing these studies and their findings because they could contribute to elaborating models for future research on green schools and their health and productivity performance outcomes. Considerations for future research are the focus of Chapter 10.

BUILDING CONDITION AND STUDENT ACHIEVEMENT

The committee identified eight studies that investigated the relationship between the overall condition of school buildings and at least two student variables. The one consistent variable was student achievement as measured by some form of standardized or normed test.

Berner (1993) investigated the relationship between parental involvement, school building condition, and student achievement in the public schools of Washington, D.C. She hypothesized that the condition of public school buildings is affected by parental involvement and that the condition of the school building in turn affects student achievement. Using a regression model that included variables for race and household income, she analyzed relationships among the size and condition of school buildings, the extent of parental involvement, and the amount of funds parents raised for the local school and compared the results with student achievement as measured using average test scores on the Comprehensive Test of Basic Skills (CTBS).

Berner found that school size, parental involvement, and building condition did have an effect on student achievement scores. The analysis indicated that student test scores increased an average of 5 percent as the condition category of school buildings improved from poor to fair condition and from fair to excellent condition. Thus, students in buildings that rated as poor had test scores that were, on average, 5 percent lower than students in school buildings categorized as fair and 10 percent lower than students in buildings categorized as excellent.

Cash (1993) investigated the relationship between certain school building conditions, student achievement, and student behavior in rural high schools in Virginia. Cash essentially used the same methodology as Berner, although in this study the condition of the building was the independent variable, and student achievement and behavior served as dependent variables. The condition of school buildings was evaluated by local school system personnel using a questionnaire which was derived from previous studies that showed a positive relationship between a particular building condition and student achievement and behavior. The factors that were looked at included air-conditioning, classroom illumination, temperature control, classroom color, graffiti, science equipment and utilities, paint schedules, roof adequacy, classroom windows, floor type, building age, supporting facilities, condition of school grounds, and furniture condition. The presence or absence of these factors or, in some cases, their quality or adequacy determined the condition category of the building: substandard, standard, and above standard.

Student achievement was measured by student test scores on the Test of Academic Proficiency (TAP), which was administered to all eleventh-graders in Virginia. The ratio of students receiving free and reduced lunches was used to control for socioeconomic status, and the Virginia Composite Index was used as a measure of local fiscal capacity, to control for the wealth of the school jurisdiction.

Cash found significant differences between the achievement scores of students in substandard buildings and those in above-standard buildings when the overall condition of the building was used as a measure. She also found that students were more affected by the cosmetic than the structural condition of a building. The difference between test scores of students in substandard and above-standard buildings ranged from 2 to 5 percentile points, depending on the subtest (i.e., mathematics, reading).

Earthman et al. (1996) used a similar methodology to conduct a study in North Dakota that included all 199 high school buildings, but used the CSTB to measure achievement. North Dakota was selected because its students traditionally score among the highest in the nation on the Scholastic Aptitude Test and the state has a relatively homogeneous, mostly rural population. Although the differences in the composite score were exactly the same as for the Cash study, there were some notable differences. The CTBS had additional subtests that the TAP did not have, such as reading, vocabulary, mathematics concepts, and spelling. In all but one subtest (social studies) of the CTBS, students in above-standard buildings outscored students in substandard buildings. The difference ranged from 1 to 9 percentile points.

Hines (1996) used the same methodology and data-gathering instrument as Cash to study large urban high schools in Virginia, and his results were basically the same: test scores for students in above-standard schools were 9 points higher for writing and science, 15 points higher for reading, and 17 points higher for mathematics compared with the same scores for students in substandard buildings.

Lanham (1999) studied the relationship between classroom conditions and student achievement in the elementary schools of Virginia that housed both third- and fifth-grade students, using the same general approach as Cash (1993). From a total of 989 elementary schools, a random sample of 299 schools was drawn. Responses were received from 197 schools, representing 66 percent participation. Lanham concluded that although certain school building and cosmetic components and features explained some of the variance in student achievement scores, the socioeconomic status of the student as represented by participation in the free and reduced-price lunch program explained most of the variance.

Schneider (2002) investigated the relationship between the condition of school buildings and student achievement scores in Washington, D.C., and Chicago, Illinois. The researcher used the reading and math scores on the Stanford Achievement Test in Washington, D.C., and the Iowa Test of Basic Skills in Chicago. After controlling for factors such as poverty, ethnicity, and school size, Schneider reported that the students in schools with good conditions were performing from 3 to 4 percentage points better on reading and math than students in buildings with poor conditions.

Lewis (2000) conducted a study based on 139 elementary, middle, and high school buildings in Milwaukee, Wisconsin. All buildings were evaluated for both condition and adequacy. The Wisconsin Student Assessment System (WSAS) was used to measure student achievement. Fourth, eighth, and tenth graders were assessed in reading, mathematics, language arts and writing, science, and social studies. Scores on these examinations were reported as a percentage of students in each school building who were achieving at or above the level "proficient." Lewis (2000, p. 11) concluded that the "significant relationships for facility measurements typically explain about 10 to 15 percent of the differences in scores across schools when the influences of the other variables were statistically controlled." When comparing student demographic indicators such as mobility rates, eligibility for free/reduced-price lunches, attendance, and suspensions, only 9 estimates out of 48 were found to be significant. Thus those indicators that were significant explained between 8 and 28 percent of the difference between test scores when other variables were controlled.

Picus et al. (2005) designed a study to examine whether higher quality buildings are related to student performance. The methodology used falls loosely into the tradition of large-scale econometric studies, which are discussed in Chapter 10. This study included approximately 300 public schools, accredited institutions, and accredited private schools in Wyoming. Building quality data were gathered in response to a court ruling related to the adequacy of the state's school funding system. Building condition scores were determined by collectively assessing up to 22 building subsystems (e.g., foundations, floors) using individual rating tools consisting of 1 to 20 questions, the answers to which were agreed on by a school representative and a subcontractor to a consultant. The consultant weighted the subsystem assessments relative to the cost of bringing the affected components into as-new condition and then averaged all the subsystem scores together to produce an overall condition score for each building.

The suitability tool purported to measure the degree to which each school was suitable for its current use, for example, whether the school was designed specifically for the grades it currently served. Ratings for suitability were self-reported by district superintendents or their designees; the report authors noted that the "suitability tool possessed a higher degree of subjectivity than the building condition instrument" (Picus et al., p. 81).

Student achievement was measured using a set of tests administered to all fourth, eighth, and eleventh graders in Wyoming (WyCAS). Three years of WyCAS results (1999-2001) were used and approximately 60,000 students were involved. The WyCAS tests comprise both multiple-choice and open-ended questions in reading, writing, and mathematics.

Two different measures of each school's achievement in each content area were used: the 3-year average of the percentage of students whose performance was "proficient" or "advanced" and the 3-year average of the scale scores. In both cases, averages for reading, writing, and mathematics were combined to arrive at an overall proxy for student achievement at a school. Correlations with building scores were computed separately for each grade level for each content area and year.

Multiple regression analyses were used to examine the relationships between building and WyCAS scores while factoring out the influence of socioeconomic status in elementary school students (as measured by the percentage of free and reduced-price lunches); there was no control for socioeconomic status in middle and high school students. After running a series of analyses, the authors found no relationship between building condition and student achievement. The "finding implies that as building condition improves, there is no likelihood that WyCAS scores will either improve or decline" (Picus et al., 2005, p. 84).

SCHOOL BUILDING FUNCTIONALITY AND STUDENT ACHIEVEMENT

Twelve studies (including Picus et al., 2005) were identified that investigated the relationship between school building functionality and student achievement. In eleven of the studies, the age of the school building was used as a surrogate for functionality. Although the age of a building might not in and of itself directly influence student achievement, an older building might not have qualities or facilities—such as thermal control, proper lighting, acoustical control, support facilities, proper laboratories, and pleasing appearance—that could affect student achievement. In the Picus study, a suitability index was calculated based on factors other than age of a building.

Using the variable of school building age, McGuffey (1982) reviewed seven studies (Thomas, 1962; Burkhead et al., 1967; Michelson, 1970; Guthrie et al., 1971; McGuffey and Brown, 1978; Plumley, 1978; and Chan, 1979). In all cases, as building age increased, student achievement decreased.

McGuffey and Brown (1978) studied 188 school districts in Georgia to explore the relationship between building age and student achievement. They used the scores on the Iowa Test of Basic Skills for fourth- and eighth-grade students and the TAP for eleventh-grade students. The statistical analyses indicated that building age could account for 0.5 percent to 2.6 percent of the variance in test scores among fourth-grade students, 0 percent to 2.6 percent of the variance among eighth-grade students, and 1.4 percent to 3.3 percent of the variance among eleventh graders.

Garrett (1981) hypothesized that when the socioeconomic status variable was statistically controlled for, the age of a facility would have a significant correlation with the achievement of students and that the achievement of students taught in unmodernized school facilities would be significantly lower than those taught in partially or fully modernized schools. When the variable for socioeconomic status was statistically controlled for, the age of the facility made a significant difference in student achievement in composition, reading, and mathematics scores on the TAP (.01). The achievement of students taught in unmodernized facilities was not significantly lower than that of those taught in partially modernized schools. However, the achievement of students taught in partially modernized schools was significantly lower than that of those taught in modern facilities.

Chan (1982) compared student attitudes toward a new school and an older school. The researcher had four hypotheses: (1) no significant difference between student attitudes toward a new building and attitudes toward an old building; (2) no difference between the attitudes of male and female students toward old and new buildings; (3) no significant difference in the attitudes of students of different races toward old and new buildings; and (4) no significant difference in attitudes between students who pay for school lunches and those who receive free and reduced-price lunches.

Chan's study used a quasi-experimental, nonequivalent control group design. The control group consisted of the 119 students in the second, third, and fourth grades in a school built around 1936. The experimental group consisted of 96 students in those same grades in a 1923-constructed building who were transferred to a new school.

After statistically adjusting the post-test scores of the control group with the corresponding pre-test scores of the experimental group, students in the experimental group scored 19 points (on a 55-point scale) higher on average than students in the control group. The difference in attitude scores was indicated by an *F*-value of 19.71, which was significant at the .0001 level. Race and socioeconomic status had no effect on student attitudes toward their school buildings. However, female students in the control group scored significantly higher than males on both pre- and posttests. All were significant at the .05 level.

Bowers and Burkett (1989) investigated the differences in student achievement, health, attendance, and behavior between two groups of students in two elementary school buildings in rural Tennessee. One school had recently been opened and was a modern building in all respects. The other building had been constructed in 1939 and had experienced few improvements to the physical structure. Two hundred eighty randomly selected fourth- and sixth-grade students in the two facilities were the subjects of the study. Principals, teachers, and socioeconomic levels of the communities were similar. The variable of age of the facility was the only major difference when comparing the achievement and behavior of the students.

Students in the modern building scored significantly higher in reading, listening, language, and arithmetic than students in the older facility (greater than .01). Discipline was needed less frequently in the new facility, even though the new school had a larger enrollment. The level of significance for analysis purposes was .01. Students in the new school building significantly outperformed students in the older building in reading, listening, language, and arithmetic. Faculty in the new building reported fewer disciplinary incidents and health issues than faculty in the old building. Attendance also was higher among students in the new building.

Phillips (1997), replicating an earlier study by Plumley (1978), found a relationship between the age of the school facility and student reading achievement scores as measured by the Iowa Test of Basic Skills and between student mathematics achievement scores and building age. The average mathematics scores for those students in new buildings increased 7.63 percentile ranks after moving into the new facility. He did not find any significant differences in attendance patterns of students enrolled in the old and new buildings.

As noted above, Picus et al. (2005) analyzed data based on a suitability tool that was purported to measure the degree to which each school building was suitable for its current use. The authors noted that this tool was more subjective than the building condition index used in the study. They found little evidence of a relationship between the suitability scores and WyCAS test scores.

LIMITATIONS OF THE CURRENT STUDIES

These studies, which found some correlation between some measure of overall building condition and student achievement, differed in several ways. Some used age as a surrogate for building condition, while others used a subjective rating of building condition. Most focused on average student achievement at the school level but a few looked at differences between individual student achievement in a modern facility considered to be in good condition and a school that was old and out of date. The studies also included urban and rural schools in several different states and in the District of Columbia. With 19 out of 20 studies showing increases in test scores for students in buildings in better condition, one might reasonably assume a relationship exists between building conditions and student achievement. In fact, the limitations of the methodologies and data used in these studies may reflect a consistent underlying bias rather than a consistent, albeit undefined, cause-and-effect relationship.

Two specific limitations lead to this conclusion. The first is the issue of omitted variable bias. None of the 20 studies included a complete set of variables that are considered to be related to student achievement. For example, parental education, family structure, family income, and teacher quality have been shown to be related to student achievement, but they were not measured. Omitting them increases the likelihood of a large positive bias for the variables that were included in the correlational or regression analysis. Thus, the coefficients on the measures of *overall* building condition are likely to have been inflated and their statistical significance is likely to have been established based on omitted variable bias.

The second significant limitation is that the relationships between overall building condition and student achievement were likely confounded because students of different socioeconomic status were not randomly assigned to schools. Minority students and students from low-income households are more likely to attend schools that are older and in substandard condition. In other words, students may not end up in old and poorly maintained schools by chance. This could give rise to reverse causality and almost certainly to ambiguity about the direction of the relationship between student achievement and overall building condition.

Additional methodological shortcomings were also present. For example, the Picus et al. (2005) study uses correlations and regression analysis or school-level averages. The current state-of-the-art research estimates multilevel models to assess not only the differences in average achievement but also the differences at the individual student level. Thus, building condition could affect the relationship between prior achievement and current achievement for individual students, meaning the students perform differently in different quality environments controlling for prior levels of achievement. This factor was not tested directly, although the study did test change in average test scores. As will be discussed in Chapter 10, the power to detect effects is severely limited in a school-level study in a small state, and the efficiency of the estimates is reduced as well.

Similarly, combining scores for different achievement domains, as was done in several studies, is a dubious practice for educational research. It is common to find that educational reforms affect math achievement more than reading achievement: By averaging math, reading, and writing together, important variation could have been missed.

Although this particular set of studies has methodological limitations, the literature on indoor environmental conditions, including lighting, noise, air quality, dampness, surface contamination, and ventilation, provides evidence that specific, as opposed to *overall*, building conditions adversely affect the indoor environments of schools and may hinder learning and impact the health of teachers and students. For example, the studies of Wargocki et al. (2005) were designed as crossover longitudinal studies intervening on specific elements (ventilation, temperature) while holding other conditions constant. These studies demonstrated modest improvements in student performance on routinely used weekly tests of verbal and math skills. Perhaps the research that attempts to relate *overall* building condition to student achievement is asking the wrong question. To understand how building conditions affect student and teacher performance, it would be better to measure one or more building performance characteristics, develop a theory linking the performance characteristics and student and/or teacher outcomes, and test the linkage using adequate measures of the outcomes of interest and fully specified regression models. Issues related to improving research are discussed in detail in Chapter 10.

CURRENT GREEN SCHOOL GUIDELINES

Current green school guidelines encourage attention to school maintenance through measures such as a computerized district maintenance plan that inventories all equipment—including HVAC, lighting, roofing, and control systems—and establishes annual tasks, with the labor and material required for their maintenance. In combination with an indoor environmental quality management plan, a computerized maintenance management plan is intended to ensure that the performance of a green school is maintained over its service life.

FINDING

Finding 8: The methodologies used in studies correlating *overall* building condition with student achievement are not adequate to determine if there is a relationship between *overall* building condition and student test scores. This research tradition seems to address a more general and diffuse question and does not produce high-quality evidence relative to either school design or specific aspects of maintenance. Improved research for understanding how specific building conditions affect student and teacher performance would measure one or more building performance characteristics, develop a theory linking those characteristics and student and/or teacher outcomes, and test the linkage using adequate measures of the outcomes of interest and fully specified regression models.

9

Processes and Practices for Planning and Maintaining Green Schools

School systems today are facing more difficult challenges than ever before. Meeting the financial challenges of building, operating, and maintaining school buildings that foster student achievement is an issue for every school district in the country. Research on a wide variety of building types over many years has demonstrated the importance of four things:

- Effective planning processes as a key ingredient to improving project success (NRC, 1989, 1990, 1994, 2000a, 2001; FFC, 1998, 2000, 2003);
- Early involvement of key stakeholders in the preproject planning process (FFC, 2000, 2003; NRC, 2000a, 2001);
- Developing a performance measurement system, including facility postoccupancy reviews, in order to understand and improve the facility delivery process (NRC, 1989, 1990, 1994, 2000a, 2001; FFC, 1998, 2000); and
- Effective training of personnel in the process and technical aspects of preproject planning and in the entire facility delivery process (FCC, 1987; NRC, 1998a, 2001).

The committee has identified additional processes that can be employed to more effectively plan, test, operate, and maintain green schools so that total life-cycle costs are decreased and total building performance is increased. Adopting a systems philosophy of design that considers the interactions of building systems and other factors, as discussed in Chapter 1, could result in schools that are more cost-effective over their entire life cycles. Building "commissioning," a quality assurance process that has proven benefits and is often cited in green school guidelines, is yet another approach.

Chapter 9 focuses on several processes that the committee believes merit inclusion in green schools guidelines: Participatory planning; building commissioning; monitoring building performance over time; post-occupancy evaluation; and training for educators and support staff.

PARTICIPATORY PLANNING

For decades, educational and community leaders have discussed the components of a successful educational program but paid scant attention to the school building as a component in the educational process. Today, there is sufficient research available to demonstrate associations between the performance of school building systems and student achievement and student and teacher health. Thus, it is clear that school facility planning amounts to more than simply ensuring the safety of bus drop-off points and specifying the required number of classrooms.

Inadequate planning for schools carries fiscal, human, and academic costs. Whether a school building is old or new, problems in design can take a significant toll. Classrooms with ambient noise can distract attention from the best-prepared lesson plans. Drab interiors, poor lighting, and the lack of pleasant social gathering spots make a school less than inviting as a place to work and learn. On the other hand, a strong facility planning process can result in significantly lower operation and maintenance costs over a facility's service life and a pleasant and healthy environment. A strong facility planning process involves asking the right questions, including a full range of stakeholders, and having a clear sense of purpose.

Research by the National Research Council (NRC), the Construction Industry Institute (CII), the Business Roundtable, and others points to the importance of conceptual or advance planning to facility acquisition and operations. It is during the planning process that the size, function, general character, special attributes, location, and budget for a school are established. Errors made at this stage can manifest in inadequate space allocations, inadequate equipment capacity, over- or undersized building systems, and so forth.

The cost influence curve (the solid curve in Figure 9.1) indicates that the ability to influence the ultimate cost of a project is greatest at the beginning, during the conceptual planning phase, and decreases rapidly as the project matures. Conversely, a project cash-flow curve (the dashed curve) indicates that conceptual planning and design costs are relatively minor and that costs escalate significantly as the project evolves through procurement, construction, and start-up phases.

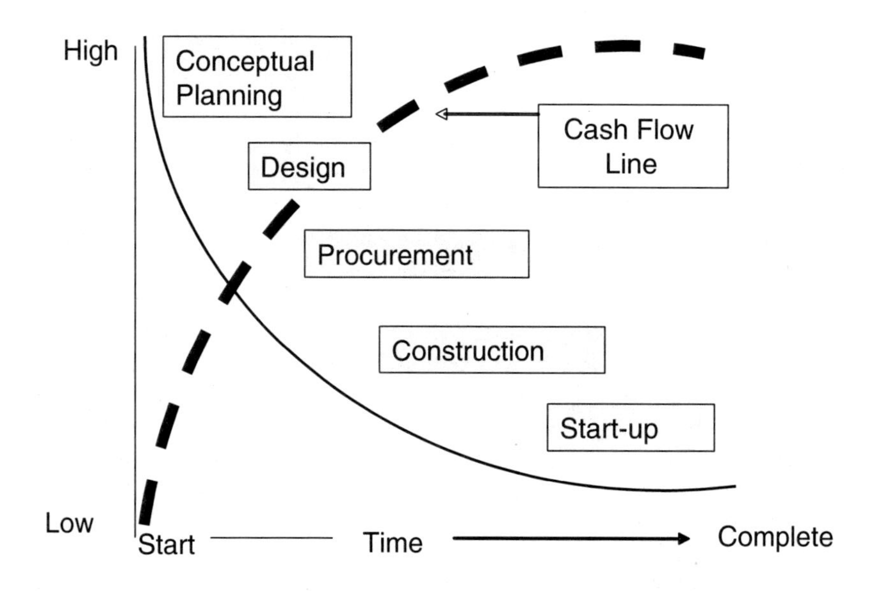

FIGURE 9.1 Cost influence and cash flow curves. SOURCE: FFC (2000).

As noted in Chapter 1, there are many stakeholders in school buildings: architects, engineers, school boards, administrators, business managers, students, teachers, staff and other building users, and elected officials. Typically, however, decisions about school facilities tend to be made by a few people who are not themselves building users, often ignoring the direct involvement of teachers and students. Involving a building committee does not by itself always solve the problem of gaining schoolwide support for a project once the design work is completed. A process that allows for face-to-face contact between users and those who influence the decisions, however, can result in a sense of ownership by all stakeholders.

Involving a full range of stakeholders in the planning of a school is challenging and places serious demands and responsibilities on all of the participants. A well-designed participatory process requires planning and management. The goal of participation is to encourage people to learn by giving them the opportunity to examine new educational environments. For this to occur, the process should be transparent, open, and encouraging of dialogue, debate, and collaboration. As more people learn about educational issues, their decisions can have positive effects on the quality of the learning environment. Participation means that parents, teachers, students, educational administrators, and public officials must engage in a dialogue about requirements and resources (Sanoff, 2001).

Sound design and planning principles should also be incorporated. Although people may voluntarily organize to participate in community projects, the technical complexity of such projects calls for professional assistance. Without guidance, community groups, teachers, administrators, and other stakeholders may respond only to crisis situations and may not achieve the overriding project goals. Thus, good planning for effective community participation requires thinking about goals and objectives, options, resources, timing, and strategies before the first public meeting. Various techniques for stakeholder participation are available, each with a different function. An evaluation of existing facilities, consisting of interviews of teachers and students followed by a walk-through evaluation by architects and engineers, can incorporate the knowledge and experience of students, staff, and teachers. This information can then be integrated into the predesign or programming stage of a new facility when building users set goals and priorities. Small group discussions can create a genuine, productive dialogue that allows people to consider many different viewpoints and to get past political disputes and will enable a thorough examination of the issues, the development of new ideas, and the establishment of common ground for constructive action. Community surveys, review boards, advisory boards, task forces, neighborhood and community meetings, public hearings, public information programs, and interactive cable television have all been used to encourage stakeholder involvement with varying degrees of success, depending on the effectiveness of the participation plan.

It is also during the planning phase for a green school when measures should be taken to ensure that the school operates as intended over its entire service life. These measures include commissioning the building(s) and monitoring building systems over time, which are discussed below.

BUILDING COMMISSIONING: QUALITY ASSURANCE FOR BUILDING PERFORMANCE

Building commissioning is a systematic quality assurance process that is often recommended in current green school guidelines. It seeks to assure before occupancy that a building performs in accordance with the stated design intent and the owner's operational needs. The commissioning of buildings is patterned after the U.S. Navy's process for commissioning its ships—namely, by taking them out for a "test drive" and subjecting them to a battery of tests so that they do not fail when they are exposed to the high seas.

The American Society of Heating, Refrigeration, and Air-Conditioning Engineers, Inc. (ASHRAE) published its original guidance document on commissioning heating, ventilation, and air-conditioning (HVAC) systems in 1989 (ASHRAE, 1989). The guidance was recently updated in *ASHRAE Guidelines 0-2005, The Commissioning Process* (ASHRAE, 2005). The guidelines call for the commissioning processes for new buildings to be "focused on verifying and documenting that the facility and all of its systems and assemblies are planned, designed, installed, tested, operated, and maintained to meet the owner's project requirements." Under the new ASHRAE guidelines, the commissioning process begins during planning and continues through occupancy.

The National Institute of Building Sciences (NIBS) serves as the secretariat for the development of a set of Guidelines for Total Building Commissioning. Table 9.1 lists the building components for which NIBS and its associated organizations have developed detailed commissioning procedures.

Other organizations such as the U.S. Green Building Council (www.usgbc.org) and the Building Commissioning Association (www.bcxa.org) have been active in developing commissioning guidance and certification for commissioning agents (those who perform the commissioning

TABLE 9.1 Building Components Included in a Comprehensive Building Commissioning Process and Relevant Organizations

Commissioning Elements	Organizations with Information
Mechanical and energy management systems	American Society of Heating, Refrigeration and Air-Conditioning Engineers
Structural systems	Structural Engineering Institute and American Society of Civil Engineers
Electrical systems	Institute of Electrical and Electronics Engineers
Lighting systems	Illuminating Engineering Society of North America
Plumbing systems	American Society of Plumbing Engineers
Fire protection systems	National Fire Protection Association
Roofing systems	National Roofing Contractors Association
Exterior envelope systems	Building Environment and Thermal Envelope Council
Elevator systems	American Society of Mechanical Engineers
Seismic protection	Building Seismic Safety Council
Telecommunications systems	Telecommunications Industry Association

process). In addition, many state, regional, and national organizations are contributing guidance for the commissioning of new and existing buildings as well as for evaluating the findings, costs, and savings of commissioning projects. The Northwest Energy Efficiency Alliance, through its "BetterBricks" initiative (www.betterbricks.org/commissioning), and the Portland Energy Conservation, Inc. (www.peci.org) offer practical experience with commissioning gained from years of working with utilities, developers, contractors, architects, and state agencies. Thus, a great deal of information is readily available on commissioning that does not have to be restated in this report. Instead, this report highlights aspects of commissioning that are critical to the process of designing, constructing, and operating green schools.

Commissioning Process

To be fully effective, a building commissioning process should begin in the school project planning phase and continue through design specifications and construction. If the commissioning process begins after planning or is restricted to specific systems. (e.g., HVAC, lighting), the potential to influence overall building performance and meet the design intent will be diminished.

As originally conceived, building commissioning is an independent function serving the owner's interest as well as the interest of the financing entity. As such, an independent commissioning agent (which may be a team of experts) serves as the owner's agent. Tseng (2005) points out that several deficiencies in current practice have led building owners to be less than enthusiastic about commissioning requirements. First, the demand for commissioning exceeds the capacity of qualified firms. Moreover, engineering and architectural firms are "overstating" their commissioning expertise. As is clear from Table 9.1, a comprehensive commissioning program requires a wide variety of expertise not often found among architectural and engineering staff. By default, the energy performance of mechanical and lighting systems became the primary focus, while systems integration, training, longer-term building performance, and technological or design innovations are not properly addressed. In some cases, the quality of commissioning has become little more than a postconstruction "checklist" evaluation. Elsewhere, owners have chosen to set up a process using firms already under contract as opposed to an independent commissioning agent. This committee believes that using an independent agent to commission a green school could result in significant savings in both short and long-term operating costs, could help to safeguard public and private financial investments, and could reduce the risk faced by a building's owners, operators, and occupants.

Special Considerations for Commissioning Green Schools

Designing and constructing a green school adds complexity to the typical facility project process. If certification as a "green" school is involved, the features that link to the basic requirements and to credits should be specified early on, in the design documents. Based on the committee's findings in Chapters 3 to 7, a commissioning process for a green school should also include the commissioning of a number of nontraditional elements. These include (1) the building envelope, to ensure that moisture management measures such as vapor barriers, flashings, and roofs are properly installed, leaks are not present, and the potential for buildup of excess moisture has been mitigated; (2) acoustical measures and systems to ensure they meet ANSI Standard 12.60; (3) indoor air quality, to ensure that ASHRAE standards for ventilation are met; and (4) lighting systems to ensure that appropriate levels of light are provided for specific tasks and spatial layouts.

The commissioning agent (a single person or a team of experts) should provide documentation for the building elements that relate to higher performance criteria, making sure these elements are not eliminated during construction or "value engineering," where specification documents are often changed to reduce costs. A commissioning agent should have a broad range of technical expertise as well as knowledge of the green building certification process. The commissioning agent should visit the site often during construction to certify that specified equipment and materials are actually being used and installed properly.

Benefits and Costs of Commissioning

The potential benefits of building commissioning include these:

- Identification of system faults/discrepancies early in the design and construction process so that they can be resolved in a timely manner while the appropriate contractual entities are still on the job.
- Documentation that the facility operates in a manner consistent with the original design intent and that green design features were certified and appropriately installed.
- Improved documentation, training, and education for operators and facility managers about specific systems and their operation.
- Reduced operation and maintenance costs.
- Fewer complaints from occupants about discomfort and, in turn, fewer service calls to building operators during the life of a building.

There is currently no standard approach to costing commissioning services. Two of the more common methods are these:

- Budgeting a percentage of the total mechanical/electrical cost of a project. A range of 2 to 6 percent is generally considered reasonable, with the higher percentages generally being utilized for those projects that are smaller or more complex.
- For accountability, setting up a separate commissioning budget that is independent of the project budget is recommended. Depending on the complexity of the building and the scope of work, commissioning costs might range between $0.10 and $0.80 per square foot.

McCarthy and Dykens (2000) state as follows: "No matter what budgeting approach is selected, it is imperative that contracts with the general and specialized contractors specify their financial liabilities. Although the commissioning agent is initially paid by the owner, additional charges incurred by the agent will be paid by the contractors if systems fail or cause delays to the schedules established for the commissioning."

MONITORING BUILDING PERFORMANCE OVER TIME

The significance of time to the performance of a school building cannot be overestimated. No matter how positive the design, engineering, and construction of a green school, ongoing, timely maintenance and repair of systems will be critical to outcomes of health, learning, and productivity: The process of creating and maintaining a green school does not end at the ribbon-cutting ceremony.

Throughout this report, emphasis has been placed on the importance of effective operation and maintenance of building systems over the 30+ years that a school building is used for educational purposes. Ensuring effective operations and maintenance, like effective building commissioning, begins during the planning phase for a new school. For example, if one of the owner's objectives is to use durable, moisture-resistant materials, this objective must be stated as a requirement during planning and transmitted as such to the designer. Similarly, if building equipment and systems are to be easily accessible for maintenance, repair, or replacement, they must be designed with that objective in mind. Once installed, materials and building systems are difficult to move and expensive to retrofit. Thus, omissions and errors in the planning phase may have repercussions for a building's performance and operating costs for 30 years or longer.

If a green school's potential health and productivity benefits are to be achieved and maintained over its service life, monitoring or diagnostic feedback about the performance of the building systems is also required.

As with other operations and maintenance features, an effective diagnostic feedback system should be considered during planning.

Building diagnostics has been defined as a set of practices that are used to assess the current performance capability of a building and to predict its likely performance over time (NRC, 1985). The essential elements of a building diagnostics program are as follows:

- Knowledge of what to measure;
- Appropriate instruments, sensors, and other tools;
- Expertise to interpret the results; and
- The capacity to predict the future condition of the building (early warning system).

Ideally, a building diagnostics program enables a school facility manager to devise corrective procedures when a potential problem is first identified to keep a small problem from becoming a much larger, more expensive one.

Building diagnostics deals with the measurement and interpretation of data and the relationship of those data to expected performance. Microprocessor technology has made it possible for manufacturers to include sensors and other intelligent controllers on heating, ventilation, and air conditioning systems, chillers, boilers, and other building systems and features. Monitoring systems are available that draw air from classrooms to a central location where it is measured for both carbon dioxide (as an indicator of ventilation effectiveness) and absolute humidity. Such systems also automatically log the data and plot and archive the results for review by the facility manager. Diagnostic systems make it possible to ensure that intended amounts of ventilation and relative humidity are provided at the time of occupancy and for many years thereafter.

Existing sensor and microprocessor technologies also have the potential to monitor and manage a range of environmental parameters that are difficult to inspect and measure during routine site visits or inspections. For example, sensor nets can be installed behind walls and bulkheads or on roofs to monitor moisture levels and indicate where moisture buildup may be occurring (NRC, 1998a). Shared sensor systems are available that can monitor carbon dioxide, absolute humidity, and carbon monoxide (to indicate the effectiveness of anti-idling measures) simultaneously at multiple locations within a building.

Diagnostic systems must be correctly operated if they are to function well over time. This, in turn, requires a workforce trained to properly operate and maintain such systems. Moreover, someone on the staff must be trained to interpret the information generated and take necessary corrective actions when a problem is indicated.

POSTOCCUPANCY EVALUATION

Postoccupancy evaluation (POE) is a method to measure and monitor building performance from the occupants' perspective. POE[1] has been defined as follows:

> A process of systematically evaluating the performance of buildings (or places) after they have been built and occupied for some time. POE differs from other evaluations of building performance in that it focuses on the requirements of building occupants, including health, safety, security, functionality and efficiency, psychological comfort, aesthetic quality, and satisfaction (FFC, 2001, p. 1).

POE is one of a number of practices aimed at understanding design criteria, predicting the effectiveness of emerging designs, reviewing completed designs, supporting building activation and facilities management, and linking user response to the performance of buildings. POE is also evolving toward more process-oriented evaluations for planning, programming, and capital asset management (FFC, 2001).

Conducting a POE a year or so after a building has been in operation has value for several reasons. First, from an energy and indoor air quality perspective, building performance needs measurement before it can be compared with benchmarked values and design specifications. Second, weather conditions challenge a building's performance. Until the structure experiences variable weather conditions, its true performance cannot be determined. Third, occupants are perhaps the best monitors of acceptability. Occupant satisfaction and their perception of cleanliness, air quality, lighting, noise, and general functionality of the space should be assessed to understand if modifications are needed and to inform design for the next building renovation or construction project. Fourth, a POE may be used to identify problems that can be further examined by technical experts, who can recommend approaches to fixing the problems. Fifth, if incorporated into the institutional memory of a school district, POEs can inform the design of future school buildings by allowing it to repeat successful features and avoid mistakes.[2]

[1]The term "facility performance evaluation" is used by some experts in the field.

[2]The National Council of Architectural Registration Boards in the United States has commissioned a monograph, *Improving Building Performance,* which contains case studies by major architectural firms in the United States that routinely conduct POEs as part of their in-house efforts to build knowledge bases and to learn from past successes and failures in building designs (Preiser, 2002).

Approaches to Successful POEs

Performance criteria for POEs for schools are based on the stated design intent and criteria contained in or inferred from the functional program. Measures include indicators related to organizational and occupant performance, such as student and teacher satisfaction and productivity and safety and security; they may also include measures of school performance as perceived by users—for example, air quality, thermal comfort, spatial comfort, ergonomics, privacy, lighting, comfort, noise, and aesthetics (Preiser et al., 1988).

Several preliminary steps are required to prepare for on-site data collection. The client should be briefed on the nature of the process, the type of activities involved, and responsibilities before a POE is conducted. Research methods and analytical techniques should be determined at this stage. In addition, background information, such as building documentation and organizational structure, is necessary, as is the identification of liaison individuals, to establish a POE plan. The plan will include the development of specific information-gathering methods, sampling methods, authorization for photographs and surveys, and data-recording sheets.

The type of POE used in a specific situation is a function of the time and resources available and the depth of knowledge necessary. Methods for collecting information include questionnaires, walk-throughs, and interviews, usually conducted with a committee representing a school's organization. Questions ordinarily focus on issues related to performance, spatial adequacy, and image. A walk-through assessment of the entire facility relies on direct observation to verify issues that may have emerged from the questionnaire. Interviews and a summary of findings often conclude the process (Sanoff, 2001). However, a forum for the discussion of outcomes can raise the awareness of the benefits of the assessment. To bring the POE process to proper completion, any commitments made to participants should be acted on.

TRAINING FOR EDUCATORS AND SUPPORT STAFF

Green schools represent a significant public investment. To protect that investment and ensure optimal building performance, facility managers, teachers, administrators, support staff, students, parents, and community partners should be informed about the design intent of a newly constructed green school. The objective is to provide all occupants with the knowledge they need to play their role in operating and maintaining a school building's environment so that it effectively supports learning and health.

The people who will be directly responsible for optimizing a building's performance are the facilities management and janitorial staff. Green school buildings potentially have sophisticated operating systems or new equipment or components that are unfamiliar to facility operators. Facilities staff, frustrated by complex lighting and ventilation systems, may find ways to override controls so that the systems do not perform as designed. Just as few parents would turn over the family car to a teenager without instruction and practice, so too should building owners give building operators and facilities staff proper instruction and training in how to properly operate and maintain the building and its systems. Similarly, janitorial staff should be trained in proper cleaning methods. For example, interrupting the transmission of infectious diseases from surfaces requires rinsing well after cleaning. This method of cleaning may be less effective if the surface is damaged or in poor condition. Results will be compromised by using a wiping cloth or material that is not initially clean or that is continually used without repeated decontamination. Training staff in proper methods of cleaning will require additional effort but should result in improved indoor environmental quality.

Helping teachers understand the physical effects of the school building and, in particular, their classrooms on their own and their students' health, well-being, comfort, and performance is critical. All too often, teachers pile books and other objects on classroom unit ventilators without understanding the potential adverse effects of blocking air flow. It is equally important to educate teachers about the effects of excessive noise on oral communication. As adults, teachers may not appreciate the additional problems that excessive noise creates for younger listeners.

Used creatively, a green school can also be a teaching tool and a means of embedding curriculum development. Essentially, a school can serve as a three-dimensional textbook (Lackney, 2004) for teaching environmental science and policy and providing hands-on examples of technologies and practices for resource conservation, recycling, energy use, air quality, and the like. In addition, a school building can be used to help teach physics, chemistry, biology, mathematics, environmental art, community building, and environmental policy.

Administrators and business managers, faced with financial constraints, may not recognize the consequences of decisions to withhold full funding for routine or preventive maintenance. Deferring maintenance for filters, drip pans, coils and other items, for example, can result in excess moisture and mold, which, in turn, can result in student and teacher illness and absenteeism. These adverse health outcomes can result in loss of funding and increased health care costs. Training that focuses on the potential long-term adverse consequences of deferred school building maintenance for occupant health and school budgets, among other things, would help to inform financial decision making.

The Environmental Protection Agency's Tools for Schools program is a well-recognized resource for training related to indoor air quality. The indoor air quality kit, for example, provides both detailed guidance and links to other information resources to help with the design of new schools as well as with the repair, renovation, and maintenance of existing facilities.

CURRENT GREEN SCHOOL GUIDELINES

Current green school guidelines do not typically address participatory planning processes or postoccupancy evaluations. They do encourage the commissioning of specific building components such as the HVAC, lighting, and energy systems. The committee believes that commissioning for green schools should be expanded to include the building envelope, acoustics, lighting, and indoor air quality. The commissioning should be done by a third-party agent before occupancy of the school to ensure that the systems meet the design goals and support the actual use of the building. Some guidelines also encourage hiring a third-party reviewer in the design and construction drawing stages to ensure that the systems can be built and maintained as designed and at the end of the warranty period to evaluate the systems.

FINDINGS AND RECOMMENDATIONS

Finding 9a: Participatory planning, commissioning, and postoccupancy evaluation are processes that can both lower building operating costs and improve performance over a building's lifetime. Current green school guidelines typically only address the practice of building commissioning.

Finding 9b: Inadequate planning for schools carries long-term fiscal, human, and academic costs. A strong planning process requires asking the right questions, involving a full range of stakeholders, and having a clear sense of purpose.

Finding 9c: A commissioning process that starts in the planning phase and continues through building occupancy can help ensure that a school building performs in accordance with the stated design criteria and the owner's operational requirements. Effective commissioning for green schools requires specific expertise in nontraditional elements such as moisture control, indoor air quality, lighting, and acoustics.

Finding 9d: If a green school's performance and potential benefits are to be maintained over its service life, building systems and features should be monitored. Such monitoring can include the use of sensors and other technologies that provide data about current indoor environmental conditions and the likely performance of a building over time.

Finding 9e: Postoccupancy evaluations can help ensure the performance of existing schools and help improve the design of future schools.

Finding 9f: Green schools represent a significant public investment. That investment can be undermined if educators, support staff, students, and other stakeholders do not have the knowledge or training to appropriately use or operate a green school.

Recommendation 9a: Future green school guidelines should stress the importance of good planning processes that allow for the effective participation of a wide range of stakeholders.

Recommendation 9b: Future green school guidelines should require for all new schools a building commissioning process that begins in planning and continues through occupancy. The commissioning agent should specifically verify that moisture-management features are properly designed and installed, that intended ventilation rates are delivered to building occupants, that the lighting system is adequately designed and installed to ensure effective lighting based on tasks and schoolroom configurations, and that acoustical measures meet the performance standards of ANSI Standard 12.60.

Recommendation 9c: Future green school guidelines should encourage the periodic monitoring of indoor environmental characteristics including moisture levels, absolute humidity, classroom temperatures, and ventilation effectiveness to ensure that performance objectives are maintained over the service life of a school.

Recommendation 9d: Educators, support staff, students, and other stakeholders should be informed of the design intent of a green school and given the appropriate information or training to fulfill their roles in using and operating a green school.

10

Linking Green Schools to Health and Productivity: Research Considerations

One of this committee's tasks was to identify avenues of research that represent potentially valuable opportunities to leverage existing knowledge into a better understanding of the relationships between green building technologies in schools and the health and performance of students and teachers. In Chapter 1, the committee recommended that

> Future green school guidelines should place greater emphasis on building systems, their interrelationships, and overall performance. Where possible, future guidelines should identify potential interactions between building systems, occupants, and operation and maintenance practices and identify conflicts that will necessitate trade-offs among building features to meet differing objectives.

In Chapters 3-7, the committee identified several avenues of research that should be pursued regarding:

- The moisture-resistance and durability of materials used in school construction as well as other properties of these materials such as generation of indoor pollutants and the environmental impacts of producing and disposing of these materials.
- Documenting a full range of costs and benefits associated with providing ventilation rates that exceed the current ASHRAE standard.
- Determining optimum temperature ranges to support student learning, teacher productivity, and occupant comfort in school buildings.

- Examining the relationships of exposures from building materials, cleaning products, and cleaning effectiveness to student and teacher health, student learning, or teacher productivity.
- The role of light on learning as well as life-long effects on health in children, particularly with regard to the role that lighting in school environments plays in regulating sleep and wakefulness in children.
- Determining optimum reverberation times that will ensure adequate speech levels without excessive reverberation in classrooms for children of various ages.
- The efficacy, costs, and benefits of alternative ventilation strategies for the dispersion and removal of airborne infectious agents.
- The use of ultraviolet germicidal irradiation in supplemental or portable air cleaning devices in school room applications and its effects on human health.

The committee recognizes that additional areas of research could also provide valuable information that could lead to improved indoor environments in schools. However, determining which areas of research may yield the most positive outcomes, and the most valuable information related to school building design, construction, and operations was beyond the resources and scope of this study.

The complexity of evaluating the effects of green schools on building occupants, the difficulty of designing research studies that can control for numerous confounding factors, and the difficulty of detecting significant effects has been discussed. As noted in Chapter 2, much of the research conducted to date has focused on one or two building systems and one or two potential outcomes. Much is still not known about the potential interactions of building systems, materials, operation and maintenance practices and their effects on building occupants, in general, or about school environments in particular. The necessary collaboration between architecture, engineering, science, medicine and social science expertise is a challenge but multidisciplinary research is required to fully study the relationships of indoor environmental quality to human health and performance outcomes.

In the following sections, the committee discusses methodologies that could potentially be used for future interdisciplinary research and identifies issues that should be addressed if the evidence base for the effects of green schools on students' and teachers' health, student learning, and teacher productivity is to be improved.

RESEARCH METHODOLOGIES

Differing methodologies and evaluation approaches generally apply to the types of outcomes being evaluated: Evaluations of health outcomes for teachers and students require a different model specification than an evaluation of student-level educational outcomes.

As detailed in Chapter 2, outcomes of educational interventions are notoriously difficult to evaluate. The two evaluation methodologies which have been given the greatest credence by the educational evaluation community are randomized experiments and econometric or regression-based techniques.

In randomized experiments students would have the same likelihood of being assigned to green schools as being assigned to conventional ones. This random selection would support an unbiased estimate of effects. Random assignment experiments allow other confounding factors to be ignored, thus greatly reducing the complexity of the evaluation. For various reasons, randomized experiments are relatively rare in educational evaluations (Shadish et al., 2002). Several innovations, such as propensity score matching have been developed to allow unbiased estimates of the effects of educational interventions (Henry et al., 2006), but, to date, these have produced less compelling evidence on the effects of interventions.

Econometric or regression-based techniques to estimate educational effects are often used in large-scale studies and place heavy data requirements on the investigation. The regressions must assess the intervention as a school input and include a full complement of variables to control for other factors. Following Boardman and Murnane (1979), Summers and Wolfe (1977), Hanushek (1997), Zimmer and Toma (2000), and Henry and Rickman (2006), the value-added specification of the school production function would be shown as:

$$Y_{1,j} = B_0 + B_1 F_{1,j} + B_1 S_{1,j} + B_1 P_{1,j} + B_1 Y_{t\text{-}1,j} + e$$

where, Y_{1j} is a measure of educational performance at time 1 for the jth student, B_0 is a constant, F_{1j} is a vector of family background influences from time $t-1$ to time t for the jth student, S_{1j} is a vector of school inputs from time $t-1$ to time t for the jth student, P_{1j} is a vector of peer influences from time $t-1$ to time t for the jth student, Y_{t-1j} is a measure of educational performance at time $t-1$ for the jth child, and e is the error term.[1]

[1]Innate ability is included in theoretical models that have been developed for estimating the school production function. However, innate ability has been routinely omitted from the reduced form equations that have been estimated largely because of the lack of clarity in defining the term and, therefore, in developing measures for it.

To evaluate the effects of green schools on educational performance, a variable indicating student attendance in a green school should be incorporated into the school inputs vector (S_{1j}). A database that includes an indicator of green school attendance would be required. In addition, if selection into green schools is a potential confounding factor, as it is likely to be, an approach that models this selection and uses two-stage, simultaneous estimation may be necessary to produce estimates that are considered unbiased by the education research community.

In similar fashion, analysis of school building characteristics using a fully specified list of components and conditions could produce unbiased estimates of the effects of those characteristics on educational performance. This would require appropriate modeling of selection into schools with particular characteristics. The studies of the effects of building characteristics currently available do not include theoretically important variables and are frequently estimated for groups of students, rather than individuals. As discussed below, these studies are likely to yield biased or inflated estimates of the effects of these building characteristics on student performance.

Other evaluation approaches could be useful in producing evidence concerning the actual effects of educational interventions such as green schools. Extended-term mixed method studies which gather information on implementation and adaptation of the reforms over time could prove useful in understanding and improving the effects of attributes of green schools on student and teacher outcomes (Chatterji, 2004). Evaluations using program theory (Weiss, 1997) or investigating mechanisms by which the program effects are produced (Mark et al., 2000; Henry and Rickman, 2006) could yield important evidence to support the causal attributions of the interventions. However, studies investigating mechanisms do not generally produce estimates of the size of an effect, primarily because they often fail to look at counterfactual data (lack of untreated or non-exposed cases). Case-control studies including crossover intervention investigations have also been used in school settings. The crossover intervention design can account for the effects of external influences (e.g., seasonal variations in respiratory infections, weather, ventilation implications) and has the advantage of having each subject serving as his or her own control. However, case control studies can only be used when the effects of the outcomes of interest are registered quickly and less likely to be useful in studies of educational outcomes where developmental age and exposure periods produce confounding interactions (maturity or treatment-maturity).

Other types of quasi-experimental study designs may be feasible for at least some future studies of the effects of green schools. Interrupted time series designs are recognized as among the strongest quasi-experimental designs. For a study on the effects of green schools, a time series of student performance before and after the introduction of a new green school could be examined. Comparison schools, where no change occurred, should be available. In addition, data may become available from green schools that are constructed and opened at various points in time. Data on relevant covariates could be examined to assess alternative explanations and strengthen statistical power.

Other quasi-experimental design features may also be possible in certain studies. For example, the Bronzaft studies on noise and learning referred to in Chapter 6 demonstrate a quasi-experimental comparison (between the train and non-train sides of the school) complemented with a removed-treatment design (after the train noise was abated). Combinations of such design features can increase confidence in causal attributions where statistical control alone might not. Similarly, the Hygge et al. (1996) study demonstrates a kind of "switching replication" design, whereby what was initially the treatment group (the schools near the airport) subsequently became the comparison group (after the airport had been moved near other schools) and vice versa.

In designing research studies to evaluate the unbiased effects of green schools on student or teacher outcomes, several additional issues must be addressed.

CONSIDERATIONS IN DESIGNING GREEN SCHOOLS-RELATED RESEARCH

Defining Green Schools for the Purpose of Scientific Inquiry

To date, green schools have been defined by their objectives and multiple design features rather than as an entity possessing a specified set of conditions common to all green schools. Considering green schools as a specific educational "intervention," and therefore an object to be evaluated, presents significant limitations. For scientific inquiry, including evaluation, an intervention must be defined in a way that is replicable and consistent across studies to produce reliable information about the effects of green schools. Therefore, a definition that accurately and completely describes green schools as an intervention is needed for the purpose of scientific evaluation.

Current green school guidelines, which rely on achieving a minimum number of design features on a checklist, could be a starting point. However, because the checklist approach permits variation in design characteristics, the "green schools intervention" will not be consistent across schools and, therefore, is likely to produce less reliable effects.

As one alternative, specific design features associated with green school designs could be evaluated. This is likely to produce more reliable and less attenuated effect size estimates but would not represent the totality of the effects, including positive effects and negative side effects that may occur within green schools.

Another alternative is to define green schools by a set of performance measures, such as ventilation and lighting. The advantage of this type of definition is to bring the definition closer to the potential learning and health effects than a design-based definition. However, it would require an additional set of measurements to be taken.

Defining Performance and Productivity Outcomes Plausibly Related to Green Schools

Two types of outcome variables were set forth in the committee's task statement: learning/productivity and health. These outcomes were expected to occur for two groups: students and teachers. Four dimensions of potential variables can be generated by a two-by-two matrix of outcomes: student learning, student health, teacher health, and teacher productivity.

Student Learning

Student learning is typically measured by student achievement on standardized tests. Green schools have the potential to affect student achievement in two distinct ways. First, better lighting and reduction of noise or other building features and characteristics may improve task performance by improving reception of the test stimuli or increasing a student's ability to concentrate on the tasks. Second, longer term improvements in indoor environmental conditions could result in greater learning and perhaps greater retention of course content.

Laboratory experiments and studies from other environments on adults show affects on task performance. However, to establish effects on learning, carefully controlled studies using randomized experiments (that include baseline measures to verify that random selection had occurred) or econometric models are needed. The period of time between baseline and post-intervention measures must be sufficient for learning effects to have occurred. In addition, the baseline should be measured in a green school (improved lighting and reduced noise environment) to control for task performance effects in the differences. Student learning could be expanded to include other types of measures of educational progress, such as graduation or promotion to higher grades. Across any of these outcomes, it would be important to use well-established, reliable, and valid measures.

Student Health

Few school-based studies include direct measures of student health. Randomized experiments with observations of student health as outcomes could be conducted for learning outcomes, with few additional complexities. Understanding how the baseline health status of students affects learning and other performance outcomes is critical to the interpretation of effects attributable to green schools. Selection bias could be important if healthier children are more likely to attend a green school and the assignment to the green school is not done at random from a pool of eligible students.

For econometric studies, a health production function must be developed to fully specify the model. As a start, replacing prior student achievement with prior student health status may be considered, but theoretical work outside the scope of this review will be needed for such studies.

Surrogate measures for student health, such as student attendance, are potentially interesting. Student attendance has the potential for reverse causality with respect to student achievement; that is, poor performance may cause absences rather than better attendance yielding higher achievement. Therefore, attendance should be used as an outcome that is intrinsically important, not as a surrogate for student health.

Teacher Health

Direct analogies can be drawn between studies of teacher health and interventions designed to improve worker health in other types of work environments. Studies could be developed to assess the impacts of green schools on teacher health using the aforementioned methodologies. However, since schools have at least an order of magnitude fewer teachers than students, fielding studies with a sufficient number of teachers to achieve an acceptable power to detect effects is likely to be an issue.

Teacher Productivity

The effect of a school building on a teacher's ability to improve student learning is principally based on the expectation that green schools result in improved teacher health which would, in turn, result in improved teacher attendance. Long-term teacher absences and the use of long-term substitutes have been associated with lower levels of student achievement in some prior studies. However, given the lack of direct effects and current controversies about measuring teacher productivity, it may be more prudent to focus on the other three outcomes, at least until measures and methods for teacher productivity are developed.

Theory Relating Cause (Green Schools) to Effects

Currently, there is no fully developed theory explaining the links between achieving a green school design and producing health and learning effects. The studies cited as evidence in this report principally address links between exposures and health or student performance. Missing are studies that empirically link green building designs with performance and studies linking building performance changes of the magnitude that can be associated with green schools to learning and health.

There is also ambiguity about the direction of the effects of specific green school design attributes. For example, having and using windows for ventilation could produce: (1) air with lower CO_2 levels; (2) higher noise levels; and (3) greater exposure to outdoor allergens. Increasing ventilation may improve the perception of air quality while also increasing background noise in a classroom. Similarly, natural lighting could produce better illumination and higher levels of glare. In each case, the consequences of a green school intervention appear to trigger mechanisms that could counteract each other in terms of the effects on educational outcomes. The potential for counterbalancing effects suggests the need for multidisciplinary evaluation teams to allow the full enumeration of plausible effects from green schools, both positive and negative, and develop intermediate outcome measures that may mediate the overall effects on learning and health.

Omitted Variable Bias

As described in Chapter 8, existing studies of the effects of overall school building condition on student learning are correlational, and do not control for selection of students into old schools or noisy schools. The studies do provide suggestive information for elaborating the models being tested to see if the effects are sustained when additional controls are added to the models. However, without fully specified econometric models or randomized experiments, omitted variable bias may be responsible for producing statistically significant effects. Evaluations using econometric models or randomized experiments would reduce the plausibility of omitted variable bias and yield estimates of the size of effects that could usefully inform decisions about green school design.

Power of the Evaluation to Detect Moderate Effects

Power refers to the ability of a study to find statistically significant effects of a particular size when those effects have in fact occurred. Because both students and teachers are nested within schools and therefore do not represent independent observations, power analysis must include the effects of the nesting, which is often referred to as clustering or cluster effects. Clustering inflates standard errors, which lowers the chances of finding statistical significance for samples of a given size. New power analysis tools (Raudenbush, 2003) indicate that as many as 50 schools could be needed to have a 75 percent probability of detecting a moderate effect size (0.30 standard deviation units), depending on the available pre-treatment covariates. Given the cost of new school construction and the complications of obtaining district cooperation to assign students to new schools at random, randomized experiments are unlikely to be both feasible and sufficiently powerful to detect moderate-sized effects on student learning or health. The power issue may motivate use of the econometric approach or more piecemeal studies of the performance features that are likely to be associated with green schools and the building performance outcomes of green schools.

Level of Aggregation

Many studies of educational effects have relied on data that are aggregated or summed to the school level rather than data measured and analyzed at the individual level. Recent research has shown many more effects when micro- or individual-level data are used (e.g., Hanushek et al., 2003). Significant measurement issues are raised by aggregating state achievement tests, which are based on individual state standards, across states. When analyzing aggregate data, it is likely that evaluations can be mounted within a single state and unlikely that in the near future there will be sufficient cases in any state for adequate power to detect effects at the school level. Strong consideration should be given to micro-level econometric studies for overall evaluations of the effects of green schools.

In light of these considerations, it may not be feasible in the near future to fund randomized experiments to evaluate the effects of green schools on health and student learning. Econometric studies with micro-level data may be pursued when there are sufficient schools within a state and data are available on student achievement and other variables to estimate fully specified models. However, it may be necessary to carry out three types of studies using various designs to justify large-scale evaluations:

1. Studies to assess the building performance characteristics that result from building designs that are constructed to meet green school design standards.
2. Studies that correlate specific building performance characteristics of the type expected from green school designs with learning and health outcomes across schools.
3. Carefully controlled efficacy studies on specific schools where baseline data and counter-factual data can be collected and, ideally, students are randomly assigned to the school from a larger population of students.

If findings from these studies are considered sufficiently positive, more costly and exacting methodologies could be justified in pursuit of unbiased estimates of the effects of green schools on health and learning.

FINDINGS

Finding 10a: Much is still not known about the potential interactions of building systems, materials, operation and maintenance practices and their effects on building occupants in general, or about school environments in particular. The necessary collaboration between architecture, engineering, physical science, medicine, and social science expertise is a challenge, but multidisciplinary research is required to fully study the potential relationship between a school building and the outcomes of students and teachers.

Finding 10b: In designing research studies to evaluate the unbiased effects of green schools on student learning or student and teacher health, several issues must be addressed. These include defining green schools for the purpose of scientific inquiry, defining performance and productivity outcomes plausibly related to green schools, and fully developing a theory explaining the links between green school design and health and learning effects. Finally, the hypotheses from these theories should be tested in ways that reduce systematic biases and provide compelling evidence about these linkages.

Finding 10c: Currently, the theory and evidence connecting green schools or characteristics associated with green schools to teacher or student outcomes is not sufficient to justify large-scale evaluations. However, the committee does consider it useful to carry out studies that assess the positive and negative consequences of the design and construction features as well as building performance characteristics that are associated with green schools using more rigorous study designs.

Finding 10d: Large-scale evaluations using randomized experiments and econometric or regression-based techniques should be conducted if they are justified from the results of smaller and less expensive studies, such as those outlined in Finding 10c. Finally, it is possible that improvements to the large-scale data sets that contain student achievement data will allow for relatively low cost studies of the effects of the school building environment on student achievement, which the committee considers to be an important side benefit.

Bibliography

AAP (American Academy of Pediatrics). 2003a. School health. Pp. 137-146 in Red Book: 2003 Report of the Committee on Infectious Diseases. L. Pickering, C. Baker, G. Overturf, and C. Prober, eds. Elk Grove Village, IL: AAP.

AAP. 2003b. Children in out-of-home child care. Pp. 123-137 in Red Book: 2003 Report of the Committee on Infectious Diseases. L. Pickering, C. Baker, G. Overturf, and C. Prober, eds. Elk Grove Village, IL: AAP.

AAP, Committee on Infectious Diseases. 2006. Recommended childhood and adolescent immunization schedule—United States, 2006. Pediatr. 117:239-240.

AASA (American Association of School Administrators). 1983. The Maintenance Gap: Deferred Repair and Renovation in the Nation's Elementary and Secondary Schools. Arlington, VA: AASA.

Achenbach, P.R. 1994. General construction principles. Pp. 283-290 in Moisture Control in Buildings. ASTM Manual Series MNL 18. H.R. Trechsel, ed. West Conshohocken, PA: American Society for Testing and Materials.

Adgate, J.L., Church, T.R., Ryan, A.D., Ramachandran, G., Fredrickson, A.L., Stock, T.H., Morandi, M.T., and Sexton, K. 2004. Outdoor, indoor, and personal exposure to VOCs in children. Environ. Health Perspect. 112:1386-1392.

Afshari, A., Matson, U., and Ekberg, L.E. 2005. Characterization of indoor sources of fine and ultrafine particles: A study conducted in a full-scale chamber. Indoor Air 15:141-150.

Anderson, K. 2004. The problem of classroom acoustics: The typical classroom soundscape is a barrier to learning. Semin. Hearing 25:117-129.

Ansari, S.A., Springthorpe, V.S., Sattar, S.A., Rivard, S., and Rahman, M. 1991. Potential role of hands in the spread of respiratory viral infections: Studies with human parainfluenza virus 3 and rhinovirus 14. J. Clin. Microbiol. 29:2115-2119.

ANSI (American National Standards Institute). 2002. Standard S12.60. Acoustical Performance Criteria, Design Requirements and Guidelines for Schools. Washington, DC: ANSI.

ANSI. 2004. Building Commissioning. Washington, DC: ANSI.

ANSI/ASHRAE (American Society of Heating, Refrigeration, and Air Conditioning Engineers). 2004. Standard 55-2004. Thermal Environmental Conditions for Human Occupancy. Washington, DC: ANSI.

ASHRAE. 1989. ASHRAE Guideline 1-1989. Commissioning of HVAC systems. Washington, DC: ASHRAE.

ASHRAE. 2004. Standard 62.1. Ventilation for Acceptable Indoor Air Quality. Washington, DC: ASHRAE.

ASHRAE. 2005. ASHRAE Guideline 0-2005. The Commissioning Process. Washington, DC: ASHRAE. Available at <www.ashrae.org/bookstore>.

ASTM (American Society of Testing and Materials). 1994. Moisture Control in Buildings. ASTM Manual Series MNL 18. H.R. Trechsel, ed. West Conshohocken, PA: American Society for Testing and Materials.

Bakó-Biró, Z., Wargocki, P., Weschler, C.J., and Fanger, P.O. 2004. Effects of pollution from personal computers on perceived air quality, SBS symptoms and productivity in offices. Indoor Air 14:178-187.

Behrentz, E., Sabin, L.D., Winer, A.M., Fitz, D.R., Pankratz, D.V., Colome, S.D., and Fruin, S.A. 2005. Relative importance of school bus-related microenvironments to children's pollutant exposure. JAWMA 55:1418-1430.

Bekö, G., Halá, O., Clausen, G., and Weschler, C.J. 2006. Initial studies of oxidation processes on filter surfaces and their impact on perceived air quality. Indoor Air 16:56-64.

Belanger, K., Beckett, W., Triche, E., Bracken, M.B., Holford, T., Ren, P., McSharry, J.E., Gold, D.R., Platts-Mills, T.A., and Leaderer, B.P. 2003. Symptoms of wheeze and persistent cough in the first year of life: Associations with indoor allergens, air contaminants and maternal history of asthma. Am. J. Epidemiol. 158:195-202.

Berner, M.M. 1993. Building conditions, parental involvement and student achievement in the D.C. public school system. Urban Education 28(1):6-29.

Berry, M.A. 2005. Educational Performance, Environmental Management, and Cleaning Effectiveness in School Environments. Available at <www.ashkingroup.com>.

Bierman, A., and K.M. Conway. 2000. Characterizing daylight photosensor system performance to help overcome market barriers. Journal of the Illuminating Engineering Society 29(1)101-105.

Blondeau, P., Iordache, V., Poupard, O., Genin, D., and Allard, F. 2005. Relationship between outdoor and indoor air quality in eight French schools. Indoor Air 15:2-12.

Bloom, B., and Tonthat, L. 2002. Summary health statistics for U.S. children: National health interview survey, 1997. Vital Health Stat. Series 10. Jan.(203):1-46.

Bloomfield, S.F. 2001. Preventing infectious disease in the domestic setting: A risk-based approach. Am. J. Infect. Control 29:207-212.

Boardman, A.E., and R.J. Murnane. 1979. Using panel data to improve estimates of the determinants of educational achievement. Sociology of Education 52:113-121.

Boubekri, M., Hull, R.B., and Boyer, L.L. 1991. Impact of window size and sunlight penetration on office workers' mood and satisfaction: A novel way of assessing sunlight. Environ. Behav. 23(4):474-493.

Bowers, J.H., and Burkett, C.W. 1989. Effects of physical and school environment on students and faculty. Educational Facility Planner 27(1): 28-29.

Boyce, P.R. 2003. Human Factors in Lighting, 2nd Edition. London and New York: Taylor and Francis.

Boyce, P.R. 2004. Reviews of Technical Reports on Daylight and Productivity. Troy, NY: Lighting Research Center. Available at <www.daylightdividends.org>.

Boyce, P.R., Hunter, C., and Howlett, O. 2003. The Benefits of Daylight Through Windows. Troy, NY: Lighting Research Center. Available at <www.daylightdividends.org>.

Bradley, J.S., and Sato, H. 2004. Speech intelligibility test results for grades 1, 3 and 6. Children in real classrooms. Paper Tu4B1.2. Proceedings of the 18th International Congress on Acoustics, Kyoto, Japan. Tokyo, Japan: Acoustical Society of Japan.

Bradley, J.S., Sato, H., and Picard, M. 2003. On the importance of early reflections for speech in rooms. J. Acoust. Soc. Am. 113(6):3233-3244.

Bradlow, A.R., Krauss, N., and Hayes, E. 2003. Speaking clearly for children with learning disabilities: Sentence perception in noise. J. Speech Lang. Hear. Res. 46(1):80-97.

Brady, M.T. 2005. Infectious disease in pediatric out-of-home child care. Am. J. Infect. Control 33:276-285.

Brady, M.T., Evans, J., and Cuartas. J. 1990. Survival and disinfection of parainfluenza viruses on environmental surfaces. Am. J. Infect. Control 18:18-23.

Brainard, G.C., Hanifin, J.P., Greeson, J.M., Byrne, B., Glickman, G., Gerner, E., and Rollag, M.D. 2001. Action spectrum for melatonin regulation in humans: Evidence for a novel circadian photoreceptor. J. Neurosci. 21(16):6405-6412.

Brennan, M., Strebel, P., George, H., Yih, W.K., Tachdjian, R., Lett, S.M., Cassiday, P., Sanden, G., and Wharton, M. 2000. Evidence for transmission of pertussis in schools, Massachusetts, 1996: Epidemiologic data supported by pulsed-field gel electrophoresis studies. J. Infect. Dis. 181:210-215.

Brickner, P.W., Vincent, R.L., Nardell, E.A., Pilek, C., Chaisson, W.T., First, M.W., Freeman, J., Wright, J.D., Rudnick, S., and Dumyahn, T. 2000. Ultraviolet upper room air disinfection for tuberculosis control: An epidemiological trial. J. Healthcare, Safety, and Infection Control 4(3):123-131.

Bronzaft, A.L. 1981. The effect of a noise abatement program on reading ability. J. Environ. Psychol. 1:215-222.

Bronzaft, A.L., and McCarthy, D.P. 1975. The effect of elevated train noise on reading ability. Environ. Behav. 7:517-528.

Bullough, J.D., and Rea, M.S. 2004. Visual performance under mesopic conditions: Consequences for roadway lighting. In Proceedings of 83rd Annual Transportation Research Board Meeting. Washington, D.C.: The National Academies Press.

Bullough, J.D., and Wolsey, R. 1998. Specifier Reports: Photosensors. Troy, NY: Rensselaer Polytechnic Institute.

Burkhead, J., Fox, T., and Holland, J.W. 1967. Input and Output in Large-City High Schools. Syracuse, NY: Syracuse University Press.

Butz, A.M., Fosarelli, P., Dick, J., Cusack, T., and Yolken, R. 1993. Prevalence of rotavirus on high-risk fomites in day-care facilities. Pediatr. 92:202-205.

Carhart, R., Tillman, T.W., and Greetis, E.S. 1969. Perceptual masking in multiple sound backgrounds. J. Acoust. Soc. Am. 45:694-703.

Carnegie Foundation. 1990. The Carnegie Foundation for Education. New York: The Carnegie Foundation.

Carskadon, M.A., Wolfson, A.R., Acebo, C., Tzischinsky, O., and Seifer, R. 1998. Adolescent sleep patterns, circadian timing, and sleepiness at a transition to early school days. Sleep 21(8):871-881.

Cash, C.S. 1993. Building condition and student achievement and behavior. Unpublished doctoral dissertation. Blacksburg, VA: Virginia Polytechnic Institute and State University.

CDC (Centers for Disease Control and Prevention). 2002. Guideline for Hand Hygiene in Health-Care Settings. Recommendations of the Healthcare Infections Control Practices Advisory Committee and the HICPAC/SHEA/APIC/IDSA Hand Hygiene Task Force MMWR Morb. Mortal. Wkly. Rep. 51:RR-16.

CDC. 2004. Asthma Prevalence, Health Care Use and Mortality, 2002. Hyattsville, MD: U.S. Department of Health and Human Services.

Chan, T.C. 1979. The impact of school building age on the achievement of eighth-grade pupils from the public schools in the state of Georgia. Unpublished doctoral dissertation. Athens, GA: University of Georgia.

Chan, T.C. 1982. A comparative study of pupil attitudes toward new and old school buildings. Greenville, SC: Greenville County School District. ERIC Document Reproduction Service No. ED 222 981.

Chatterji, M. 2004. Evidence on "what works": An argument for extended-term mixed-method (ETMM) evaluation designs. Educational Researcher 33(9): 3-13.

CHPS (Collaborative for High Performance Schools). 2004. Best Practices Manual. Volumes 1-4. Available at <www.chps.net>.

CHPS. 2005. Web site at <www.chps.net>.

Christian, J.E. 1994. Moisture sources. Pp. 176-182 in Moisture Control in Buildings. ASTM Manual Series MNL 18. H.R. Trechsel, ed. West Conshohocken, PA: American Society for Testing and Materials.

CII (Construction Industry Institute). 1986. Evaluation of Design Effectiveness. Report RS8-1. Austin, TX: University of Texas, Austin.

Clausen, G. 2004. Ventilation filters and indoor air quality: A review of research from the International Centre for Indoor Environment and Energy. Indoor Air 14(Suppl 7):202-207.

Cohen, S., Evans, G.W., Krantz, D.S., and Stokols, S. 1980. Physiological, motivational and cognitive effects of aircraft noise on children: Moving from the laboratory to the field. Am. Psychol. 35:231-243.

Collins, B.L. 1975. Windows and People: A Literature Survey: Psychological Reaction to Environments with and Without Windows. National Bureau of Standards Building Science Series 70. Washington, DC: U.S. Department of Commerce.

Cotton, K. 2001. New Small Learning Communities: Findings From Recent Literature. Portland, OR: Northwest Regional Educational Laboratory. Available at <www.nwrel.org/scpd/sirs/nslc.pdf>.

Council of Great City Schools. 1987. Report on Deferred Maintenance to the Council of the Great City Schools. Washington, DC: Council of Great City Schools.

Cox, C.S. 1987. The Aerobiological Pathway of Microorganisms. Chichester, NY: Wiley.

Cox-Ganser, J.M., White, S.K, Jones, R., Hilsbos, K., Storey, E., Enright, P.L., Rao, C.Y., and Kreiss, K. 2005. Respiratory morbidity in office workers in a water damaged building. Environ. Health Perspect. 113:485-490.

Crump, D., Squire, R., Brown, V., Yu, C., Coward, S., and Aizlewood, C. 2005. Investigation of volatile organic compounds in the indoor air of a school over a one year period following refurbishment. Pp. 659-663 in Proceedings of Indoor Air 2005. Beijing, China: Tsinghua University Press.

Cuttle, K. 1983. People and windows in workplaces. Pp. 203-212 in Proceedings of the People and Physical Environment Research Conference. Wellington, New Zealand: New Zealand Ministry of Works and Development.

Daisey, J.M., Angell, W.J., and Apte, M.G. 2003. Indoor air quality, ventilation, and health symptoms in schools: An analysis of existing information. Indoor Air 13:53-64.

Darling-Hammond, L. 1999. Teacher Quality and Student Achievement: A Review of State Policy Evidence. Seattle, WA: University of Washington.

Davison, K.L., Andrews, N., White, J.M., Ramsay, M.E., Crowcroft, N.S., Rushdy, A.A., Kaczmarski, E.B., Monk, P.N., and Stuart, J.M. 2004. Clusters of meningococcal disease in school and preschool settings in England and Wales: What is the risk? Arch. Dis. Child. 89:256-260.

Deal, B.M., Rose, W., and Riley, S.E. 1998. Commissioning for humidified buildings. USACERL Technical Report 99/03. Champaign, IL: U.S. Army Corps of Engineers Construction Engineering Research Laboratories.

Demos, G.D., Davis, S., and Zuwaylif, F.F. 1967. Controlled physical environments. Building Research 4:60-62.

Dettenkofer, M., Wenzler, S., Amthor, S., Antes, G., Motschall, E., and Daschner, F.D. 2004. Does disinfection of environmental surfaces influence nosocomial infection rates? A systematic review. Am. J. Infect. Control 32:84-89.

Dowell, S.F., Simmerman, J.M., Erdman, D.D., Wu, J.S. Chaovavanich, A., Javadi, M., Yang, J.Y., Anderson, L.J., Tong, S., and Ho, M.S. 2004. Severe acute respiratory syndrome coronavirus on hospital surfaces. Clin. Infect. Dis. 39:652-657.

Duguid, J. 1946. The size and duration of air-carriage of respiratory droplets and droplet-nuclei. J. Hygiene 44:471-479.

Early, E., Battle, K., Cantwell, E., English, J., Lavin, J.E., and Larson, E. 1998. Effect of several interventions on the frequency of handwashing among elementary public school children. Am. J. Infect. Control 26:263-269.

Earthman, G.I., Cash, C.S., and Van Berkum, D. 1996. Student achievement and behavior and school building condition. J. School Business Management 8(3):26-37.

Eccles, R. 2000. Spread of common colds and influenza. Available at <www. ifh-homehygiene.org/newspage/new05.htm>.

Educational Writers'Association. 1989. Wolves at the Schoolhouse Door: An Investigation of the Conditions of Public School Buildings. Washington, DC: Educational Writers' Association.

Edward, D., Elford, W., and Laidlaw, P. 1943a. Studies of air-borne virus infections. I. Experimental technique and preliminary observations on influenza and infectious ectromelia. J. Hyg. (Lond) 43(1):1-10.

Edward, D., Elford, W., and Laidlaw, P. 1943b. Studies of air-borne virus infections: II. The killing of virus aerosols by ultra-violet radiation. J. Hyg. (Lond) 43(1):11-15.

Edwards, D.A., Man, J.C., Brand, P., Katstra, J.P., Sommerer, K., Stone, H.A., Nardell, E., and Scheuch, G. 2004. Inhaling to mitigate exhaled bioaerosols. Proc. Natl. Acad. Sci. U.S.A. 101(50):17383-17388.

Elliott, L., Connors, S., Kille, E., Levin, S., Ball, K., and Katz, D. 1979. Children's understanding of monosyllabic nouns in quiet and in noise. J. Acoust. Soc. Am. 66:12-21.

Elliott, L.L. 1979. Performance of children aged 9 to 17 years on a test of speech intelligibility in noise using sentence material with controlled word predictability. J. Acoust. Soc. Am. 66:651-653.

Engvall, K., Wickmann, P., and Norbäck, D. 2005. Sick building syndrome and perceived indoor environment in relation to energy saving by reduced ventilation flow during heating season: A 1 year intervention study in dwellings. Indoor Air 15:120-126.

Evans, G.W., and Maxwell, L. 1997. Chronic noise exposure and reading deficits. Environ. Behav. 29:638-656.

Evans, H.S., and Maguire, H. 1996. Outbreaks of infectious intestinal disease in schools and nurseries in England and Whales 1992 to 1994. Commun. Dis. Rep. CDR Rev. 6(7): R103-R108.

Fabian, M.P., McDevitt, J. and Milton, D.K. 2006. Modes of transmission of respiratory viral infections. In Exacerbations of Asthma. O'Byrne, P., and Johnston, S.L., eds. London: Taylor and Francis. In press.

Fang, L., Clausen, G., and Fanger, P.O. 1999a. Impact of temperature and humidity on the perception of indoor air quality. Indoor Air 8:80-90.

Fang, L., Clausen, G., and Fanger, P.O. 1999b. Impact of temperature and humidity on the perception of indoor air quality during immediate and longer whole-body exposures. Indoor Air 8:276-284.

Fanger, P.O. 2000a. Perceived air quality and ventilation requirements. Chapter 22 in Indoor Air Quality Handbook. Spengler, J.D. Samet, J.M., and McCarthy, J.F., eds. New York: McGraw Hill.

Fanger, P.O. 2000b. Indoor air quality in the 21st century: The search for excellence. Indoor Air 10:68-73.

FCC (Federal Construction Council). 1987. Quality Control on Federal Projects. Washington, DC: National Academy Press.

Federspiel, C.C., Fisk, W.J., Price, P.N., Liu, G., Faulkner, D., Dibartolomeo, D.L., Sullivan, D.P., and Lahiff, M. 2004. Worker performance and ventilation in a call center: Analyses of work performance data for registered nurses. Indoor Air 14(Suppl 8):41-50.

Fetveit, A., Skjerve, A., and Bjorvatn, B. 2003. Bright light treatment improves sleep in institutionalized elderly—An open trial. Int. J. Geriatr. Psychiatry 18(6):520-526.

FFC (Federal Facilities Council). 1998. Government/Industry Forum on Capital Facilities and Core Competencies. Washington, DC: National Academy Press.

FFC. 2000. Adding Value to the Facility Acquisition Process: Best Practices for Reviewing Facility Designs. Washington, DC: National Academy Press.

FFC. 2001. Learning from Our Buildings: A State-of-the Practice Summary of Post-Occupancy Evaluation. Washington, DC: National Academy Press.

FFC. 2003. Starting Smart: Key Practices for Developing Scopes of Work for Facility Projects. Washington, DC: The National Academies Press.

FFC. 2005. Implementing Health-Protective Features and Practices in Buildings. Washington, DC: The National Academies Press.

Figueiro, M.G., and Rea, M.S. 2005. LEDs: Improving the sleep quality of older adults. In Proceedings of the CIE Midterm Meeting and International Lighting Congress. Leon, Spain, May 18-21. Vienna, Austria: Commission Internationale de l'Eclairage.

Finitzo-Hieber, T., and Tillman, T.W. 1978. Room acoustics effects on monosyllabic word discrimination ability for normal and hearing-impaired children. J. Speech Hear. Res. 21:440-458.

First, M., Nardell, E., Chaisson, W., and Riley R. 1999a. Guidelines for the application of upper-room ultraviolet germicidal irradiation for preventing transmission of airborne contagion—Part I: Basic principles. ASHRAE Transactions Symposia 105(1):869-876.

First, M., Nardell, E., Chaisson, W., and Riley, R. 1999b. Guidelines for the application of upper-room ultraviolet germicidal irradiation for preventing transmission of airborne contagion—Part II: Design and operational guidance. ASHRAE Transactions Symposia 105(1): 877-887.

First, M.W., Weker, R.A., Yasui, S., and Nardell, E.A. 2005. Monitoring human exposures to upper-room germicidal ultraviolet irradiation. J. Occup. Environ. Hyg. 2(5):285-292.

Fischer, J.C., and Bayer, C.W. 2003. Humidity control in school facilities. ASHRAE Journal 45:30-37. Available at <doas-radiant.psu.edu/Fischer_Article_on_School_IAQ_03.pdf>.

Fisk, W.J. 2000. Health and productivity gains from better indoor environment and their relationship with building energy efficiency. Annual Review of Energy and the Environment 25:537-566.

Fisk, W.J., Faulkner, D., Palonen, J., and Seppänen, O. 2002. Performance and costs of particle air filtration technologies. Indoor Air 12(4):223-234.

Fisk, W.J., Seppänen, O., Faulkner, D., and Huang, J. 2003. Economizer system cost effectiveness: Accounting for the influence of ventilation rate on sick leave. Pp. 361-367 in Healthy Buildings 2003: Energy-Efficient Healthy Buildings. Proceedings of ISIAQ 7th International Conference Healthy Buildings, December 7-12, Singapore. Volume 3. K.W. Tham, C. Sekhar, D. Cheong, eds. Espoo, Finland: International Society of Indoor Air Quality and Climate.

Franke, D.L., Cole, E.C., Leese, K.E., Foarde, K.K., and Berry, M.A. 1997. Cleaning for improved indoor air quality: An initial assessment for effectiveness. Indoor Air 7:41-54.

Fritzell, B. 1996. Voice disorders and occupations. Logoped. Phoniatr. Vocol. 21:7-12.

GAO (General Accounting Office). 1995a. School Facilities: Condition of America's Schools. GAO/HEHS-95-61.Washington, DC: GAO.

GAO. 1995b. School Facilities: America's Schools Not Designed or Equipped for 21st Century. GAO/HEHS-955-95. Washington, DC: GAO.

GAO. 1996. School Facilities: America's Schools Report Differing Conditions, Report to Congress. GAO/HEHS-96-103. Washington, DC: GAO.

Garrett, D.M. 1981. The impact of school building age on the academic achievement of high school pupils in the state of Georgia. Unpublished doctoral dissertation. Athens, GA: University of Georgia.

Gerone, P., Couch, R., Keefer, G., Douglas, R., Derrenbacher, E., and Knight, V. 1966. Assessment of experimental and natural viral aerosols. Bacteriol. Rev. **30**(3):576-588.

Girou, E., Loyeau, S., Legrand, P., Oppein, F., and Brun-Buisson, C. 2002. Efficacy of handrubbing with alcohol based solution versus standard handwashing with antiseptic soap: Randomised clinical trial. Br. Med. J. 325:362.

Glickman, G., Hanifin, J.P., Rollag, M.D., Wang, J., Cooper, H., and Brainard, G.C. 2003. Inferior retinal light exposure is more effective than superior retinal exposure in suppressing melatonin in humans. J. Biol. Rhythms 18(1):71-79.

Goldmann, D.A. 2001. Epidemiology and prevention of pediatric viral respiratory infections in health-care institutions. Emerg. Infect. Dis. 7:249-253.

Goodland, J. 1983. A Place Called School. Prospects for the Future. New York: McGraw-Hill.

Gotaas, C., and Starr, C.C. 1993. Vocal fatigue among teachers. Folia Phoniatr (Basel) 45:120-129.

Graudenz, G.S., Oiveira, C.H., Tribess, A., Mendes, C., Jr., Latorre, M.R., and Kalil, J. 2005. Association of air-conditioning with respiratory symptoms in office workers in tropical climate. Indoor Air 15:62-66.

Green, C.F., and Scarpino, P.V. 2002. The use of ultraviolet germicidal irradiation (UVGI) in disinfection of airborne bacteria. Environ. Eng. Policy 3:101-107.

Green, K.B., Pasternack, B.S., and Shore, R.E. 1982. Effects of aircraft noise on reading ability of school-age children. Arch. Environ. Health 37:24-31.

Guinan, M.E., McGuckin-Guinan, M., and Sevareid, A. 1997. Who washes hands after using the bathroom? Am. J. Infect. Control 25:424-425.

Guinan, M., McGuckin, M., and Ali, Y. 2002. The effect of a comprehensive handwashing program on absenteeism in elementary schools. Am. J. Infect. Control 30:217-220.

Guthrie, J.W., Kleindorfer, G.B., Levin, H.M., and Stout, R.T. 1971. Schools and Inequity. Cambridge, MA: MIT Press.

Gwaltney, J.M., and Hendley, J.O. 1982. Transmission of experimental rhinovirus infection by contaminated surfaces. Am J. Epidemiol 116: 828-833.

Gwaltney, J.M., Moskalski, P.B., and Hendley, J.O. 1978. Hand to hand transmission of rhinovirus colds. Ann. Int. Med. 88:463-467.

Haines, M.M., Stansfeld, S.A., Brentnall, S., Head, J., Berry, B., Jiggins, M., Hygge, S. 2001a. The West London School Study: The effects of chronic aircraft noise exposure on child health. Psychol. Med. 3:1385-1396.

Haines, M.M., Stansfeld, S.A., Job, R.F.S., Berglund, B., and Hea, J. 2001b. Chronic noise exposure, stress responses, mental health and cognitive performance in schools. Psychol. Med. 31:265-277.

Hall, C.B. 1981. Nosocomial viral respiratory infections: Perennial weeds on pediatric wards. Am. J. Med. 70:670-676.

Hall, C.B. 2000. Nosocomial respiratory syncytial virus infections: The "cold war" has not ended. Clin. Infect. Dis. 31:590-596.

Hall, C.B. 2001. Respiratory syncytial virus and parainfluenza virus. N. Engl. J. Med. 344:1917-1928.

Hall, C.B., and Douglas, R.G., Jr. 1981. Modes of transmission of respiratory syncytial virus. J. Pediatr. 99:100-103.

Hall, C.B., Douglas, R.G., Jr., Schnabel, K.C., and Geiman, J.M. 1981. Infectivity of respiratory syncytial virus by various routes of inoculation. Infect. Immun. 33:779-783.

Hall, C.B., Geiman, J.M., and Douglas, R.G., Jr. 1980. Possible transmission by fomites of respiratory syncytial virus. J. Infect. Dis. 141:98-102.

Hammond, B., Ali, Y., Fendler, E., Dolan, M., and Donovan, S. 2000. Effect of hand sanitizer use on elementary school absenteeism. Am. J. Infect. Control 28:340-346.

Hansen, H.L., and Hanssen, S.O. 2002. Education, indoor environment and HVAC solutions in school buildings–Consequences of differences in paradigm shifts. Pp. 800-806 in Proceedings of Indoor Air 2002, the Ninth International Conference on Indoor Air Quality and Climate, Monterey, CA, 2002. Espoo, Finland: International Society of Indoor Air Quality and Climate. Available at <www.chps.net/info/iaq_papers/PaperVI.3.pdf>.

Hanssen, S.O. 2004. HVAC–The importance of clean intake section and dry filter in cold climate. Indoor Air 14(Suppl 7):195-201.

Hanushek, E.A. 1997. Assessing the effects of school resources on student performance: An update. Educational Evaluation and Policy Analysis 21:143-163.

Hanushek, E.A., Kain, J.F., Markman, J.M., and Rivkin, S.G. 2003. Does peer ability affect student achievement? J. Appl. Econometrics 18:527-544.

Harley, D., Harrower, B., Lyon, M., and Dick, A. 2001. A primary school outbreak of pharyngoconjunctival fever caused by adenovirus type 3. Commun. Dis. Intell. 25:9-12.

Harper, G.J. 1961. Airborne micro-organisms: Survival tests with four viruses. J. Hygiene (Lond.) **59**:479-486.

Hartman, G. 1946. The effects of noise on children. J. Educ. Psychol. 37:149-161.

Heerwagen, J., and Heerwagen, D. 1986. Lighting and psychological comfort. Lighting Design and Application 6:47-51.

Henderson, H.I. 2003. Understanding the dehumidification performance of air conditioning equipment as part-load condition: Background and Theory. Unpublished Seminar 40 at ASHRAE 2003 Annual Meeting, July 2, 2003, Kansas City, MO.

Hendley, J.O., Wenzel, R.P, and Gwaltney, J.M. 1973. Transmission of rhinovirus colds by self-inoculation. N. Engl. J. Med. 288(26):1361-1364.

Henry, G.T., and Rickman, D.S. 2006. Do peers influence children's skill development in preschool? Economics of Education Review. In press.

Henry, G.T., Gordon, C.S., and Rickman, D.S. 2006. Early Education Policy Alternatives: Comparing the Quality and Outcomes of Head Start and Pre-Kindergarten. Educational Evaluation and Policy Analysis. In press.

Heschong-Mahone Group. 1999. Daylighting in Schools: An Investigation Into the Relationship Between Daylighting and Human Performance. Fair Oaks, CA: Heschong-Mahone Group.

Heschong-Mahone Group. 2001. Daylighting in Schools: Reanalysis Report. Technical Report P500-03-082-A-3. Sacramento, CA: California Energy Commission.

Heschong-Mahone Group. 2003. Windows and Classrooms: A Study of Student Performance and the Indoor Environment. P500-03-082-A-7. Fair Oaks, CA: Heschong-Mahone Group.

Heuer, H., Kleinsorge, T., Klein, W., and Kohlisch, O. 2004. Total sleep deprivation increases the costs of shifting between simple cognitive tasks. Acta Psychol. (Amst.) 117(1):29-64.

Higgins, A.T., and Turnure, J.E. 1984. Distractability and concentration of attention in children's development. Child Dev. 55:1799-1810.

Hines, E.C. 1996. Building condition and student achievement and behavior. Unpublished doctoral dissertation. Blacksburg, VA: Virginia Polytechnic Institute and State University.

Hodgson, A.T., Shendell, D.G., Fisk, W.J., and Apte, M.G. 2004. Comparison of predicted and derived measures of volatile organic compounds inside four new relocatable classrooms. Indoor Air 14(Suppl 8):135-144.

Hodgson, M., and Nosal, E.M. 2002. Effect of noise and occupancy on optimal reverberation times for speech intelligibility in classrooms. J. Acoust. Soc. Am. 111(2):931-939.

Hygge, S. 2003. Classroom experiments of the effects of different noise sources and sound levels on long term recall and recognition in children. Appl. Cogn. Psychol. 17:895-914.

Hygge, S., Evans, G.W., and Bullinger, M. 1996. The Munich Airport noise study: Cognitive effects on children from before to after the change of airports. Pp. 2189-2194 in Proceedings of Internoise '96. Volume 5. Poughkeepsie, NY: Institute of Noise Control Engineering.

IFH (International Scientific Forum on Home Hygiene). 2002a. Hygiene procedures in the home and their effectiveness: A review of the scientific evidence base. Available at <www.ifh-homehygiene.org>.

IFH. 2002b. The infection potential in the domestic setting and the role of hygiene practice in reducing infection. Available at <www.ifh-homehygiene.org>.

IFH. 2002c. Recommendations for selections of suitable hygiene procedures for use in the domestic environment. Available at <www.ifh-homehygiene.org>.

IOM (Institute of Medicine). 1991. Adverse Effects of Pertussis and Rubella Vaccines. Washington, DC: National Academy Press.

IOM. 1993. Adverse Events Associated with Childhood Vaccines: Evidence Bearing on Causality. Stratton, K.R., Howe, C.J., and Johnston, R.B., eds. Washington, DC: National Academy Press.

IOM. 1994. Veterans and Agent Orange: Health Effects of Herbicides Used in Vietnam. Washington, DC: National Academy Press.

IOM. 1996. Veterans and Agent Orange. Update 1996. Washington, DC: National Academy Press.

IOM. 1999. Veterans and Agent Orange. Update 1998. Washington, DC: National Academy Press.

IOM. 2000. Clearing the Air: Asthma and Indoor Air Exposures. Washington, DC: National Academy Press.

IOM. 2003. Veterans and Agent Orange. Update 2002. Washington, DC: The National Academies Press.

IOM. 2004. Damp Indoor Spaces and Health. Washington, DC: The National Academies Press.

Jaakola, M.S., Nordman, H., Piipari, R., Uitti, J., Laitinen, J., and Karjalainen, A. 2002. Indoor dampness and molds and development of adult-onset asthma: A population-based incident study. Environ. Health Perspect. 110:543-547.

Jackson, G.J., and Holmes, J.G. 1973. Let's keep it simple—What we want from daylight. Light and Lighting 12:80-82.

Jennings, J.R., Monk, T.H., and van der Molen, M.W. 2003. Sleep deprivation influences some but not all processes of supervisory attention. Psychol. Sci. 14(5):473-479.

Jennings, L.C., and Dick, E.C. 1987. Transmission and control of rhinovirus colds. Eur. J. Epidemiol. 3(4):327-35.

Jensen, S.L. 1963. Carotenoids.X. On the constitution of minor carotenoids of Rhodopseudomonas Lp518. Aeta Chem. Scand. 17:303-312.

Jewett, M., Rimmer, D.W., Duffy, J.F., Klerman, E.B., Kronauer, R., and Czeisler, C.A. 1997. Human circadian pacemaker is sensitive to light throughout subjective day without evidence of transients. Am. J. Physiol. 273:R1800-R1809.

Johnston, S.L., Pattemore, P.K., Sanderson, G., Smith, S., Campbell, M.J., Josephs, L.K., Cunningham, A., Robinson, B.S., Myint, S.H., Ward, M.E., Tyrrell, D.A., and Holgate, S.T. 1996. The relationship between upper respiratory infections and hospital admissions for asthma: a time-trend analysis. Am. J. Respir. Crit. Care Med. 54(3 Pt 1):654-660.

Johnston, S.L., Pattemore, P.K., Sanderson, G., Smith, S., Lampe, F., Josephs, L., Symington, P., O'Toole, S., Myint, S.H., Tyrrell, D.A.J., and Holgate, S.T. 1995. Community study of role of viral infections in exacerbations of asthma in 9-11 year old children. Br. Med. J. 310(6989):1225-1229.

Junqua, J.C. 1996. The influence of acoustics on speech production: A noise-induced stress phenomenon known as the Lombard reflex. Speech Commun. 20:13-22.

Karlberg, A., Magnusson, K., and Nilsson, U. 1992. Air oxidation of d-limonene (the citrus solvent) creates potent allergens. Contact Dermatitis 26:332-340.

Kemper, A.R., Bruckman, D., and Freed, G.L. 2004. Prevalence and distribution of corrective lenses among school-age children. Optom. Vis. Sci. 81(1):7-10.

Khetsuriani, N., Bisgard, K., Prevots, D.R., Brennan, M., Wharton, M., Pandya, S., Poppe, A., Flora, K., Dameron, G., and Quinlisk, P. 2001. Pertussis outbreak in an elementary school with high vaccination coverage. Pediatr. Infect. Dis. J 20:1108-1112.

Kimel, L.S. 1996. Handwashing education can decrease illness absenteeism. J. Sch. Nurs. 12:14-16, 18.

Knudsen, V.O., and Harris, C. 1950. Acoustical Designing in Architecture. Hoboken, N.J.: John Wiley and Sons.

Kuller, R., and Lindsten, C. 1992. Health and behavior of children in classrooms with and without windows. J. Environ. Psychol. 12:305-317.

Kwok, A.G. 2000. Thermal Comfort Concepts and Guidelines. Chapter 15 in Indoor Air Quality Handbook. Spengler, J.D., Samet, J.M., McCarthy, J.F., eds. New York: McGraw Hill.

Lackney, J.A. 1999. Assessing school facilities for learning. Starkville, MS: Mississippi State University Educational Design Institute.

Lackney, J.A. 2004. Twelve design principles based on brain-based learning research. DesignShare: The International Forum for Innovative Schools. Available at <www.designshare.com/research/brainbasedlearn98.htm>.

Lagercrantz, L., Wistrand, M., Willen, U., Wargocki, P., Witterseh, T., and Sundell, J. 2000. Negative impact of air pollution on productivity: Previous Danish findings repeated in new Swedish test room. Pp. 653-658 in Healthy Buildings 2000. Vol. 1. Espoo, Finland: International Society of Indoor Air Quality and Climate.

Lanham III, J.W. 1999. Relating building and classroom conditions to student achievement in Virginia's elementary schools. Unpublished doctoral dissertation. Blacksburg, VA: Virginia Polytechnic Institute and State University.

Larson, C.T. 1965. The effect of Windowless Classrooms on Elementary School Children. Ann Arbor, MI: Architectural Research Laboratory, Department of Architecture, University of Michigan.

Leadership News. 1994. Performance-based funds to finance educational reforms. August 15, p. 4.

Lee, B.R., Feaver, S.L., Miller, C.A., Hedberg, C.W., and Ehresmann, K.R. 2004. An elementary school outbreak of varicella attributed to vaccine failure: Policy implications. J. Infect. Dis. 190:477-483.

Lee, G.M., Salomon, J.A., Friedman, J.F., Hibberd, P.L., Ross-Degnan, D., Zasloff, E., Bediako, S., and Goldmann, D.A. 2005. Illness transmission in the home: A possible role for alcohol-based hand gels. Pediatr. 115:852-860.

Leslie, R., and Hartleb, S. 1990. Windows, variability, and human response. Proceedings of the International Daylighting Conference. Moscow. Moscow, Russia: Research Institute of Building Physics (NIISF).

Leslie, R., and Hartleb, S. 1991. Some effects of the sequential experience of windows on human response. J. Illum. Eng. Soc. (Winter):91-99.

Levin, H. 1997. Systematic Evaluation and Assessment of Building Environmental Performance. Second International Conference on Buildings and the Environment, CIB TG8, Environmental Assessment of Buildings, June 9-12, Paris. Rotterdam, The Netherlands: International Council for Research and Innovation in Building and Construction.

Lewis, M.G. 2000. Facility Condition and Student Test Performance in the Milwaukee Public Schools. ED 459 593. Scottsdale, Ariz.: Council of Educational Facility Planners, International.

Li, Y., Duan, S., Yu, I.T., and Wong, T.W. 2005. Multi-zone modeling of probable SARS virus transmission by airflow between flats in Block E, Amoy Gardens. Indoor Air **15**(2):96-111.

Lstiburek, J., and Carmody, J. 1994. The Moisture Control Handbook: Principles and Practices for Residential and Small Commercial Buildings. New York: Van Nostrand Reinhold.

Ludlow, A.M. 1976. The functions of windows in buildings. Lighting Res. Technol. 6(2):57-68.

MacDowell, A.L., and Bacharier, L.B. 2005. Infectious triggers of asthma. Immunol. Allergy Clin. North Am. **25**(1):45-66.

Maier, W.C., Arrighi, H.M., Murray, B., Llewellyn, and Redding, G.C. 1997. Indoor risk factors for asthma and wheezing among school children. Environ. Health Perspect. 105:208-214.

Maniccia, D., Rutledge, B., Rea, M., and Narendran, N. 1998. A Field Study of Lighting Controls. Troy, NY: Lighting Research Center, Rensselaer Polytechnic Institute.

Mannino, D., Homa, D., Akinbami, L., Moorman, J., Gwynn, C., and Redd, S. 2002. Surveillance for asthma prevalence—United States 1980-1999. MMWR Surveillance Summaries 51(SS01):1-13.

Marchisio, P., Gironi, S., Esposito, S., Schito, G.C., Mannelli, S., Principi, N., and Ascanius Project Collaborative Group. 2001. Seasonal variations in nasopharyngeal carriage of respiratory pathogens in healthy Italian children attending day-care centres or schools. J. Med. Microbiol. 50:1095-1099.

Mark, M.M., Henry, G.T., and Julnes, G. 2000. Evaluation: An Integrated Framework for Understanding, Guiding, and Improving Policies and Programs. San Francisco: Jossey-Bass.

Markus, T.A. 1965. The function of windows: A reappraisal. Building Science 2:97-121.

Marshall, N.B. 1987. The effects of different signal-to-noise ratios on the speech recognition scores of children. Ph.D. thesis. Tuscaloosa, AL: University of Alabama.

Master, D., Hess Longe, S.H., and Dickson, H. 1997. Scheduled hand washing in an elementary school population. Fam. Med. 29:336-339.

Mbithi, J.N., Springthorpe, V.S., Boulet, J.R., and Sattar, S.A. 1992. Survival of hepatitis A virus on human hands and its transfer on contact with animate and inanimate surfaces. J. Clin. Microbiol. 30:757-763.

McCarthy, J.F., and Dykens, M.J. 2000. Building commissioning for mechanical systems. Chapter 61 from Indoor Air Quality Handbook. J.D. Spengler, J.M. Samet, and J.F. McCarthy, eds. New York: McGraw-Hill Publishers.

McGuffey, C.W. 1982. Facilities. Chapter 10, pp. 237-288 in Improving Educational Standards and Productivity. W. Herbert, ed. Berkeley, CA: McCutchan Publishing Corp.

McGuffey, C.W., and Brown, C.L. 1978. The impact of school building age on school achievement in Georgia. CEFP J. 16:6-9.

McIntosh, K. 2005. Coronaviruses in the limelight. J. Infec. Dis. 191:489-491.

McIntyre, I.M., Norman, T.R., Burrows, G.D., and Armstrong, S.M. 1989a. Human melatonin suppression by light is intensity dependent. J. Pineal Res. 6(2):149-156.

McIntyre, I.M., Norman, T.R., Burrows, G.D., and Armstrong, S.M. 1989b. Quantal melatonin suppression by exposure to low intensity light in man. Life Sci. 45(4):327-332.

McLean, R.L. 1961. The effect of ultraviolet radiation upon the transmission of epidemic influenza in long-term hospital patients. Am. Rev. Respir. Dis. **83**(2 Part 2):36-38.

Meadows, E., and Le Saux, N. 2004. A systematic review of the effectiveness of antimicrobial rinse-free hand sanitizers for prevention of illness-related absenteeism in elementary school children. BMC Public Health 4:50.

Melikov, A., Pitchurov, G., Naydenov, K., and Langkilde, G. 2005. Field study on occupant comfort and the office thermal environment in rooms with displacement ventilation. Indoor Air 15:205-214.

Mendell, M.J., and Heath, G.A. 2004. Do indoor pollutants and thermal conditions in schools influence student performance? A critical review of the literature. Indoor Air 15:27-52.

Mendell, M.J., Fisk, W.J., Kreiss, K., Levin, H., Alexander, D., Cain, W.S., Girman, J.R., Hines, C.J., Jensen, P.A., Milton, D.K., Rexroat, L.P., and Wallingford, K.M. 2002. Improving the health of workers in indoor environments: priority research needs for a national occupational research agenda. Am. J. Public Health 92(9):1430-1440.

Meng, Q.Y., Turpin, B.J., Korn, L, Weisel, CP, Morandi, M, Colome, S., Zhang, J.J., Stock, T., Spektor, D., Winer, A., Zhang, L., Lee, J.H., Giovanetti, R., Cui, W., Kwon, J., Alimokhtari, S., Shendell, D., Jones, J., Farrar, C., and Maberti, S. 2005. Influence of ambient (outdoor) sources on residential indoor and personal PM2.5 concentrations: analyses of RIOPA data. J. Expo. Anal. Environ. Epidemiol. 15(1):17-28.

Menzies, D., Pasztor, J., Nunes, F., Leduc, J., and Chan, C.H. 1997. Effect of a new ventilation system on health and well-being of office workers. Arch. Environ. Health 52:360-367.

Menzies, D., Popa, J., Hanley, J.A., Rand, T., and Milton, D.K. 2003. Effect of ultraviolet germicidal lights installed in office ventilation systems on workers' health and well-being: Double-blind multiple crossover trial. Lancet 362:1785-1791.

Michelson, S. 1970. The Association of Teacher Resourcefulness with Children's Characteristics. Do Teachers Make a Difference? A Report on Recent Research on Pupil Achievement. U.S. Office of Education Report OE-58042. Washington, DC: U.S. Government Printing Office.

Middleton, D. 1993. Upper respiratory tract infections of childhood. Part I. The common cold. Farm. Pract. Recertification 15:60ff.

Milton, K., Glenross, P., and Walters, M. 2000. Risk of sick leave associated with outdoor air supply rate, humidification, and occupant complaint. Int. J. Indoor Air Qual. Clim. 10:211-221.

Moglia, D., Smith, A., MacIntosh, D.L., and Somers, J.L. 2006. Prevalence and implementation of IAQ programs in U.S. Schools. Environ. Health Perspect. 114:141-146.

Moncur, J.P., and Dirks, D. 1967. Binaural and monaural speech intelligibility in reverberation. J. Speech Hear. Res. 10:186-195.

Moore, R.Y. 1997. Circadian rhythms: Basic neurobiology and clinical applications. Annu. Rev. Med. 48:253-266.

Moser, M.R., Bender, T.R., Margolis, H.S., Noble, G.R., Kendal, A.P., and Ritter, D.G. 1979. An outbreak of influenza aboard a commercial airliner. Am. J. Epidemiol. **110**(1):1-6.

Musher, D.M. 2003. How contagious are common respiratory tract infections? N. Engl. J. Med. 348:1256-1266.

Myatt, T.A., Johnston, S.L., Zuo, Z., Wand, M., Kebadze, T., Rudnick, S., and Milton, D.K. 2004. Detection of airborne rhinovirus and its relation to outdoor air supply in office environments. Am. J. Respir. Crit. Care Med. **169**(11):1187-90.

Náběelek, A., and Donahue, A. 1984. Perception of consonants in reverberation by native and non-native listeners. J. Acoust. Soc. Am. 75:632-634.

Náběelek, A.K., and Pickett, J.M. 1974. Reception of consonants in a classroom as affected by monaural and binaural listening, noise, reverberation, and hearing aids. J. Acoust. Soc. Am. 56(2):628-639.

Náběelek, A.K., and Robinson, P.K. 1982. Monaural and Binaural speech perception in reverberation for listeners of various ages. J. Acoust. Soc. Am. 71(5):1242-1248.

Nardell, E.A., Keegan, J., Cheney, S.A., and Etkind, S.C. 1991. Airborne infection. Theoretical limits of protection achievable by building ventilation. Am. Rev. Respir. Dis. 144(2):302-306.

National Sleep Foundation. 2005. Sleep in America Poll. Available at <www.sleepfoundation.org/_content/hottopics/2005_summary_of_findings.pdf>.

Nazaroff, W.W., and Weschler, C.J. 2004. Cleaning products and air fresheners: Exposure to primary and secondary air pollutants. Atmos. Environ. 38:2841-2865.

NCES (National Center for Education Statistics). 2000. Condition of America's Public School Facilities: 1999. NCES2000-032. Washington, DC: U.S. Department of Education.

NCES. 2002. Digest of Educational Statistics, 2001. NCES2002130. Washington, DC: U.S. Department of Education.

Ne'eman, E., and Longmore, J. 1973. Physical aspects of windows: Integration of daylight with artificial light. Proceedings of CIE Conference on Windows and Their Function in Architectural Design. Istanbul. Vienna, Austria: Commission Internationale de l'Eclairage.

NEA (National Education Association). 2000. Modernizing Our Schools: What Will It Cost? Washington, DC: NEA.

Neely, A.N., and Orloff, M.M. 2001. Survival of some medically important fungi on hospital fabrics and plastics. J. Clin. Microbiol. 39:3360-3361.

Nel, A. 2005. Air pollution related illness: Effects of particles. Science 308(5723):804-807.

Neuman, A., and Hochberg, I. 1983. Children's perception of speech in reverberation. J. Acoust. Soc. Am. 73:2145-2149.

New York DDC (New York Department of Design and Construction). 1999. High Performance Building Guidelines. New York: DDC.

Nilsson, A., Kihlström, E., Lagesson, V., Wessén, B., Szponar, B., Larsson, L., and Tagesson, C. 2004. Microorganisms and volatile organic compounds in airborne dust in damp residences. Indoor Air 14:74-82.

Nörback, D., Walinder, R., Wieslander, G., Smedje, G., Erwall, C., and Venge, P. 2000. Indoor air pollutants in schools: Nasal patency and biomarkers in nasal lavage. Allergy 55:163-170.

NORDAMP (Nordic Interdisciplinary Review of the Scientific Evidence on Associations Between Exposure to "Dampness" in Buildings and Health Effects). 2002. Indoor Air 15(10):7-16.

Nördquist, B., and Jensen, L. 2005. Analysis of fan assisted natural ventilation in schools. In Proceedings of Indoor Air 2005. Beijing, China: Tsinghua University Press.

NRC (National Research Council). 1985. Building Diagnostics. Washington, DC: National Academy Press.

NRC. 1989. Improving the Design Quality of Federal Buildings. Washington, DC: National Academy Press.

NRC. 1990. Achieving Designs to Budget for Federal Facilities. Washington, DC: National Academy Press.

NRC. 1994. On the Responsibilities of Architects and Engineers and Their Clients in Federal Facilities Development. Washington, DC: National Academy Press.

NRC. 1998a. Stewardship of Federal Facilities: A Proactive Strategy for Managing the Nation's Public Assets. Washington, DC: National Academy Press.

NRC. 1998b. Assessing the Need for Independent Project Reviews in the Department of Energy. Washington, DC: National Academy Press.

NRC. 2000a. Improving Project Management in the Department of Energy. Washington, DC: National Academy Press.

NRC. 2000b. Outsourcing Management Functions for the Acquisition of Federal Facilities. Washington, DC: National Academy Press.

NRC. 2001. Progress in Improving Project Management at the Department of Energy. Washington, DC: National Academy Press.

Nuanualsuwan, S., and Cliver, D.O. 2003. Capsid functions of inactivated human picornaviruses and feline calicivirus. Appl. Environ. Microbiol. 69(1):350-357.

Nuanualsuwan, S., Mariam, T., Himathokan, S. and Cliver, D.O. 2002. Ultraviolet inactivation of feline calcivirius, human enteric viruses, and coliphages. Photochem. Photobiol. 76:406-410.

OFEE (Office of Federal Environmental Executive). 2003. The Federal Commitment to Green Building: Experiences and Expectations. Available at <www.ofee.gov/sb/fgb_report.pdf>.

Ohlund, L.S., and Ericsson, K.B. 1994. Elementary school achievement and absence due to illness. J. Genet. Psychol. 155:409-421.

ORNL (Oak Ridge National Laboratory). 1988. The Building Foundation Design Handbook. ORNL/SUB/86-724143/1. Oak Ridge, TN: ORNL.

Otsu, R. 1998. A mass outbreak of gastroenteritis associated with group C rotaviral infection in schoolchildren. Comp. Immunol. Microbiol. Infect. Dis. 21:75-80.

Park, K.H., and Jo, W.K. 2004. Personal volatile organic compound (VOC) exposure of children attending elementary schools adjacent to industrial complex. Atmos. Environ. 38:1303-1312.

Parrish, J.A. 2005. Immunosuppression, skin cancer, and ultraviolet A radiation. N. Engl. J. Med. 353:2712-2713.

Phillips, R.W. 1997. Educational facility age and the academic achievement and attendance of upper elementary school students. Unpublished doctoral dissertation. Athens, GA: University of Georgia.

Phipatanakul, W., Cronin, B., Wood, R.A., Eggleston, P.A., Shih, M.C., Song, L., Tachdjian, R., and Oettgen, H.C. 2004. Effect of environmental intervention on mouse allergen levels in homes of inner-city Boston children with asthma. Ann. Allergy Asthma Immunol. 92(4):420-425.

Picard, M., and Bradley, J.S. 2001. Revisiting speech interference in classrooms. Audiology 40:221-244.

Picus, L.O., Marion, S.F., Calvo, N., and Glenn, W.J. 2005. Understanding the relationship between student achievement and the quality of educational facilities: Evidence from Wyoming. Peabody J. Educ. 80(3):71-95.

Piette, M., and Nordman, B. 1996. Cost and benefits from utility-funded commissioning of energy efficient measures in 16 buildings. ASHRAE Trans. Symposia 10291:482-491.

Plumley, J.P., Jr. 1978. The impact of school building age on the academic achievement of pupils from selected schools in the state of Georgia. Unpublished doctoral dissertation. Athens, GA: University of Georgia.

Porter, D.R., ed. 2000. The Practice of Sustainable Development. Washington, DC: Urban Land Institute.

Powell, F.J. 1994. Climate. Pp. 91-175 in Moisture Control in Buildings. ASTM Manual Series MNL 18. Heinz R. Trechsel, ed. West Conshohocken, PA: American Society of Testing and Materials.

Preciado, J.A., Garcia Tapia, R., and Infante, J.C. 1998. Prevalence of voice disorders among educational professionals. Factors contributing to their appearance or their persistence. Acta Otorrinolaringology Espana 49(2):137-142.

Preiser, W.F.E. 2002. Improving Building Performance. NCARB Monograph Series. Washington, DC: National Council of Architectural Registration Boards. Available at <www.ncarb.org>.

Preiser, W.F.E, Rabinowitz, H.Z., and White, E.T. 1988. Post-Occupancy Evaluation. New York: Van Nostrand Reinhold.

Raudenbush, S.W. 2003. Designing field trials of educational innovations. Available at <http://www.ssicentral.com/hlm/resources.html>.

Raudenbush, S.W., and Bryk, A.S. 2002. Hierarchical Linear Models: Applications and Data Analysis Methods. Thousand Oaks, CA: Sage Publications.

Raymond, J., and Aujard, Y. 2000. Nosocomial infections in pediatric patients: A European, multicenter prospective study. Infect. Control Hosp. Epidemiol. 21:260-263.

Raywid, M.A. 1996. Taking Stock: The Movement to Create Mini-Schools, Schools-Within-Schools, and Separate Small Schools. New York: ERIC Clearinghouse on Urban Education.

Rea, M.S., ed. 2000. IESNA Lighting Handbook: Reference and Application. 9th edition. New York: Illuminating Engineering Society of North America.

Rea, M.S., and Bullough, J.D. 2001. Application efficacy. J. Illum. Eng. Soc. 30(2):73-96.

Rea, M.S., and Ouellette, M.J. 1991. Relative visual performance: A basis for application. Light. Res. Technol. 23(3):135-144.

Rea, M.S., Bullough, J.D., Freyssinier-Nova, J.P., and Bierman, A. 2004. A proposed unified system of photometry. Light. Res. Technol. 36(2):85-111(27).

Rea, M.S., Figueiro, M.G., and Bullough, J.D. 2002. Circadian photobiology: An emerging framework for lighting practice and research. Light. Res. Technol. 34(3):177-190.

Rea, M.S., Ouellette, M.J., and Kennedy, M.E. 1985. Lighting and task parameters affecting posture, performance, and subjective ratings. J. Illum. Eng. Soc. 15(1):231-238.

Reich, R., and Bradley, J.S. 1998. Optimizing classroom acoustics using computer model studies. Can. Acoust. 26(3):15-21.

Reichert, T.A., Sugaya, N., Fedson, D.S., Glezen, W.P., Simonsen, L., and Tashiro, M. 2001. The Japanese experience with vaccinating schoolchildren against influenza. N. Engl. J. Med. **344**(12):889-896.

Reichman, J., and Healy, W.C. 1983. Learning disabilities and conductive hearing loss involving otitis media. J. Learn. Disabil. 16:272-278.

Reid, K.J., and Zee, P.C. 2004. Circadian rhythm disorders. Semin. Neurol. 24(3):315-325.

Reitzig, M., Mohr, S., Heinzow, B., and Knoppel, H. 1998. VOC emissions after building renovations: traditional and less common indoor air contaminants, potential sources, and reported health complaints. Indoor Air 8:91-102.

Remington, P.L., Hall, W.N., and Davis, I.H. 1985. Airborne transmission of measles in a physician's office. JAMA 253:1574-1571.

Ribéron, J., O'Kelly, P., Maupetit, F., and Robine, E. 2002. Indoor air quality in schools: The impact of ventilation conditions and indoor activities. Pp. 109-114 in Proceedings of Indoor Air 2002, the Ninth International Conference on Indoor Air Quality and Climate, Monterey, CA, 2002. Espoo, Finland: International Society of Indoor Air Quality and Climate.

Riley, E.C., Murphy, G., and Riley, R.L. 1978. Airborne spread of measles in a suburban elementary school. Am. J. Epidemiol. **107**(5):421-432.

Riley, R.L., and Nardell, E.A. 1989. Clearing the air. The theory and application of ultraviolet air disinfection. Am. Rev. Respir. Dis. **139**(5):1286-1294.

Riley, R.L., and O'Grady, F. 1961. Airborne Infection: Transmission and Control. New York: Macmillan.

Roberts, L., Smith, W., Jorm, L., Patel, M., Douglas, R.M., and McGilchrist, C. 2000. Effect of infection control measures on the frequency of upper respiratory infection in child care: A randomized, controlled trial. Pediatr. 105:738-742.

Robinson, J. 2001. Infectious diseases in schools and child care facilities. Pediatr. Rev. 22:39-46.

Rochon-Edouard, S., Pons, J.L., Veber, B., Larkin, M., Vassal, S., and Lemeland, J.F. 2004. Comparative in vitro and in vivo study of nine alcohol-based handrubs. Am. J. Infect. Control 32:200-204.

Rogers, M., Weinstock, D.M., Eagan, J., Kiehn, T., Armstrong, D., and Sepkowitz, K.A. 2000. Rotavirus outbreak on a pediatric oncology floor: Possible association with toys. Am. J. Infect. Control 28:378-380.

Rosen, L.N., Targum, S.D., Terman, M., Bryant, M.J., Hoffman, H., Kasper, S.F., Hamovit, J.R., Docherty, J.P., Welch, B., and Rosenthal, N.E. 1990. Prevalence of seasonal affective disorder at four latitudes. Psychiatry Res. 31(2):131-144.

Rosenthal, N. 1998. Winter Blues. Seasonal Affective Disorder. What It Is and How to Overcome It. New York and London: Guilford Press.

Rosenthal, N., Levendosky, A.A., Skwerer, R.G., Joseph-Vanderpool, Jr., Kelly, K.A., Harbin, T., Kasper, S., Dellabella, P. and Wehr, T.A. 1990. Effects of light treatment on core body temperature in seasonal affective disorder. Biol. Psychiatry 27(1):39-50.

Rosenthal, N., Sack, D., Parry, B., Mendelsom, W., Tamarkin, L., and Wehr, T. 1985. Seasonal affective disorder and phototherapy. Ann. N.Y. Acad. Sci. 453:260.

Rousseau, M.Z. 2003. Heat, air and moisture control strategies for managing condensation in walls. Pp.1-11 in Building Science Insight Seminar Series 2003 Proceedings, 15 Cities across Canada. NRCC-46734. Ottawa, Ontario: National Research Council Canada.

Roy, C.J., and Milton, D.K. 2004. Airborne transmission of communicable infection—The elusive pathway. N. Engl. J. Med. **350**(17):1710-1712.

Rudnick, S.N., and Milton, D.K. 2003. Risk of indoor airborne infection transmission estimated from carbon dioxide concentration. Indoor Air **13**(3):237-245.

Rusak, B., Eskes, G.A., and Shaw, S.R. 1997. Lighting and Human Health: A Review of the Literature. Ottawa: Canada Mortgage and Housing Corporation.

Rutala, W.A., and Weber, D.J. 2004. Disinfection and sterilization in health care facilities: What clinicians need to know. Clin. Infect. Dis. 39:702-709.

Rutala, W.A., Barbee, S.L., Aguiar, N.C., Sobsey, M.D., and Weber, D.J. 2000. Antimicrobial activity of home disinfectants and natural products against potential human pathogens. Infect. Control Hosp. Epidemiol. 21:33-38.

Samet, J.M., Cushing, A.H., Lambert, W.E., Hunt, W.C., McLaren, L.C., Young, S.A., and Skipper, B.J. 1993. Comparability of parent reports of respiratory illnesses with clinical diagnoses in infants. Am. Rev. Respir. Dis. 148:441-446.

Sandora, T.J., Taveras, E.M., Shih, M.C., Resnick, E.A., Lee, G.M., Ross-Degnan, D., and Goldmann, D.A. 2005. A randomized, controlled trial of a multifaceted intervention including alcohol-based hand sanitizer and hand-hygiene education to reduce illness transmission in the home. Pediatr. 116:587-594.

Sanoff, H. 2001. Community Participation Methods in Design and Planning. New York: John Wiley and Sons.

Satlin, A., Volicer, L., Ross, V., Herz, L., and Campbell, S. 1992. Bright light treatment of behavioral and sleep disturbances in patients with Alzheimer's disease. Am. J. Psychiatry 149:1028-1032.

Schneider, M. 2002. Public school facilities and teaching: Washington, D.C., and Chicago. Available at <www.edfacilities.org>.

Schwartz, J. 2004. Air pollution and children's health. Pediatrics 113:1037-1043.

Scott, E. 2001. The potential benefits of infection control measures in the home. Am. J. Infect. Control 29:247-249.

Seppänen, O.A. 1999. Estimated cost of indoor climate in Finnish buildings. Pp. 13-18 in Proceedings of Indoor Air 1999. The 8th International Conference on Indoor Air Quality and Climate, Edinburgh, Scotland, August 8-13, 1999. G.J. Raw, C.E. Aizlewood, and P.R Warren, eds. Espoo, Finland: International Society of Indoor Air Quality and Climate.

Seppänen, O.A., and Fisk, W.J. 2002. Association of ventilation system type with SBS symptom in office workers. Indoor Air 12:98-112.

Seppänen, O.A., and Fisk, W.J. 2004a. A Model to Estimate the Cost Effectiveness of Indoor Environment Improvements in Office Work. Berkeley, CA: Lawrence Berkeley Laboratory.

Seppänen, O.A., and Fisk, W.J. 2004b. Summary of human responses to ventilation. LBNL-55748. Available at <http://respositories.cdlib.org/lbnl/LBNL-55748>.

Seppänen, O.A., and Fisk, W.J. 2005. Some quantitative relations between indoor environmental quality and work performance or health. Proceedings of Indoor Air 2005, 10th International Conference on Indoor Air Quality and Climate, Beijing, China. Beijing, China: Tsinghua University Press.

Seppänen, O., and Vuolle, M. 2000. Cost effectiveness of some remedial measures to control summer time temperatures in an office building. Pp. 665-670 in Proceedings of Healthy Buildings 2000, Volume 1. Espoo, Finland: International Society of Indoor Air Quality and Climate.

Seppänen, O.A., Fisk, W.J., and Mendell, M.J. 1999. Association of ventilation rates and CO_2 concentrations with health and other human responses in commercial and institutional buildings. Indoor Air 9:226-252.

Shadish, W., Cook, T., and Campbell, D. 2002. Experimental and Quasi-experimental Designs for Generalized Causal Inferences. Boston, MA: Houghton-Mifflin Company.

Shendell, D.G., Barnett, C., and Boese, S. 2004a. Science-based recommendations to prevent or reduce potential exposure to biological, chemical, and physical agents in schools. J. School Health 74(10):390-396.

Shendell, D.G., Prill, R., Fisk, W.J., Apte, M.G., Blake, O., and Faulkner, D. 2004b. Associations between classrooms' CO_2 concentrations and student attendance in Washington and Idaho. Indoor Air 14:333-341.

Shirey, D.B. 2003. Understanding the dehumidification performance of air conditioning equipment at part-load: Test Results. Unpublished Seminar at ASHRAE 2003 Annual Meeting, July 2, 2003, Kansas City, MO.

Singer, B.C., Destaillats, H., Hodgson, A.T., and Nazaroff, W.W. 2006. Cleaning products and air fresheners: emissions and resulting concentrations of glycol ethers and terpenoids. Indoor Air 16:179-191.

Singer, B.C., Hodgson, A.T., Hotchi, T., and Kim, J.J. 2004. Passive measurement of nitrogen oxides to assess traffic-related pollutant exposure for the East Bay Children's Respiratory Health Study. Atmos. Environ. 38:393-403.

Slezak, J.A., Persky, V.W., Kuiz, F.J., and Byers, C. 1998. Asthma prevalence and risk factors in selected Head Start sites in Chicago, Ill. J. Asthma 35:203-212.

Smedje, G., and Norbäck, D. 2000. New ventilation systems at select schools in Sweden—Effects on asthma and exposure. Arch. Environ. Health 55:18-25.

Smedje, G., and Norbäck, D. 2001. Irritants and allergens at school in relation to furnishings and cleaning. Indoor Air 11:127-133.

Smith, E., Lemke, J., Taylor, M., Kirchner, H.L., and Hoffman, H. 1998. Frequency of voice problems among teachers and other occupations. J. Voice 12:480-488.

Stansfeld, S.A., Berglund, B., Clark, C., Lopez-Barrio, I., Fischer, P., Ohrstrom, E., Haines, M.M., Hygge, S., van Kamp, I., and Berry, B.F. 2005. Aircraft and road traffic noise and children's cognition and health: A cross-national study. Lancet 365:1942-1949.

Stark, P.C., Burge, H.A., Ryan, L.M., Milton, D.K., and Gold, D.R. 2003. Fungal levels in the home and lower respiratory tract illnesses in the first year of life. Am. J. Resp. Crit. Care Med. 168:232-237.

State of Maine. 1887. 33rd Annual Report of the State Superintendent of Common Schools, State of Maine, 1886. Augusta, ME: Sprague and Sons, Printers to the State.

Straube, J.F., 2002. Moisture, materials, and buildings. HPAC Engineering Magazine. April.

Straube, John F. 2002. Air barriers role in preserving IAQ. ASHRAE IAQ Applications, Technical Feature, Spring.

Summers, A.A., and Wolfe, B.L. 1977. Do schools make a difference? Amer. Econ. Rev. 67(4):639-652.

Takaro, T.K., Krieger, J.W., and Song, L. 2004. Effect of environmental interventions to reduce exposure to asthma triggers in homes of low-income children in Seattle. J. Exposure Analysis Environ. Epidemiol. 14:S133-S143.

Taskinen, T., Hyvarinen, A., Meklin, T., Husman, T., Nevalainen, A., and Korppi, M. 1999. Asthma and respiratory infections in school children with special reference to moisture and mold problems in the school. Acta Paediatr. 88:1373-1379.

Tham, K.W., William, H.C., Sekhar, S.C., Wyon, D.P., Wargocki, P., Fanger, P.O. 2003. The SBS-symptom and environmental perceptions of office workers in the Tropics at two air temperatures and two ventilation rates. Pp.182-187 in Proceedings of Healthy Buildings 2003, Singapore, Vol. 1. Espoo, Finland: International Society of Indoor Air Quality and Climate.

Thapan, K., Arendt, J., and Skene, D.J. 2001. An action spectrum for melatonin suppression: Evidence for a novel non-rod, non-cone photoreceptor system in humans. J. Physiol. 535:261-267.

Thomas, J.A. 1962. Efficiency in education: A study of the relationship between selected inputs and mean test scores in a sample of senior high schools. Unpublished doctoral dissertation. Palo Alto, CA: Stanford University.

Thorn, A. 1998. Building-related health problems: Reflections on different symptom prevalence among pupils and teachers. Int. J. Circumpolar Health 57:249-256.

TIAX. 2003. Matching the Sensible Heat Ratio of Air Conditioning Equipment with the Building Load Heat Ratio. Cambridge, MA: TIAX LLC.

Titze, I.R., Lemke, J., and Montequin, D. 1996. Populations in the U.S. workforce who reply on voice as a primary tool of trade. NCVS Status Progress Rep. 10:127-132.

Toftum, J., Jorgensen, A.S., and Fanger, P.O. 1998. Upper limits for air humidity to prevent warm respiratory discomfort. Energy and Buildings 28:15-23.

Tseng, P.C. 2005. Commissioning sustainable buildings. ASHRAE J. 47:S20-S24.

Tuomainen, M., Smolander, J., Korhonen, P., Eskola, L., and Seppänen, O. 2003. Potential economic benefits of balancing air flows in an office building. Pp. 516-521 in Healthy Buildings 2003: Energy-Efficient Healthy Buildings. Proceedings of ISIAQ 7th International Conference Healthy Buildings, December 7-12, Singapore. Volume 2. K.W. Tham, C. Sekhar, D. Cheong, eds. Espoo, Finland: International Society of Indoor Air Quality and Climate.

Turek, F.W., and Zee, P.C. 1999. Introduction to sleep and circadian rhythms. Pp. 1-17 in Regulation of Sleep and Circadian Rhythm. Turek, F.W., and Zee, P.C., eds. New York: Marcel Dekker.

Turner, R.B., Biedermann, K.A., Morgan, J.M., Keswick, B., Ertel, K.D., and Barker, M.F. 2004. Efficacy of organic acids in hand cleansers for prevention of rhinovirus infections. Antimicrob. Agents Chemother. 48:2595-2598.

Tye, R.P. 1994. Relevant Moisture Properties of Building Construction Materials. Pp. 35-53 in Moisture Control in Buildings. ASTM Manual Series MNL 18. H.R. Trechsel, ed. West Conshohocken, PA: American Society for Testing and Materials.

USGBC (United States Green Building Council). 2005. LEED-NC Reference Guide, Version 2.2. Available at <www.usgbc.org>.

van Someren, E.J., Kessler, A., Mirmirann, M., and Swaab, D.F. 1997. Indirect bright light improves circadian rest-activity rhythm disturbances in demented patients. Biol. Psychiatry 41:955-963.

Vernon, S., Lundblad, B., and Hellstrom, A.L. 2003. Children's experiences of school toilets present a risk to their physical and psychological health. Child Care Health Dev. 29:47-53.

Vital Health and Statistics. 1998. Vital Health and Statistics, Current Estimates from the National Health Interview Survey, 1995. Atlanta, GA: U.S. Centers for Disease Control and Prevention and the National Center for Health Statistics.

Wallace, L.A. 1991. Comparison of risks from outdoor and indoor exposures to toxic chemicals. Environ. Health Perspect. 95:7-13.

Wang, D., Federspiel, C.C., and Arens, E. 2005. Correlation between temperature satisfaction and unsolicited complaint rates in commercial buildings. Indoor Air 15:13-18.

Wargocki, P. 2003. Estimate of economic benefits from investment in improved indoor air quality in office building. Pp. 383-387in Healthy Buildings 2003: Energy-Efficient Healthy Buildings. Proceedings of ISIAQ 7th International Conference Healthy Buildings, December 7-12, Singapore. Volume 3. K.W. Tham, C. Sekhar, D. Cheong, eds. Espoo, Finland: International Society of Indoor Air Quality and Climate.

Wargocki, P., Sundell, J., Bischof, J., Brundrett, G., Fanger, P.O., Gyntelberg, F., Hanssen, O., Harrison, P., Pickering, A., Seppänen, O.A., and Wouters, P. 2002. Ventilation and health in non-industrial indoor environments: Report from European Multidisciplinary Scientific Consensus Meeting (EUROVEN). Indoor Air 12:113-128.

Wargocki, P., Wyon, D.P., and Sundell, J. 2000. The effects of outdoor air supply rate in an office on perceived air quality, Sick Building Syndrome (SBS) symptoms and productivity. Indoor Air 10(4):222-236.

Wargocki, P., Wyon, D.P., Baik, Y.K., Clausen, G., and Fanger, P.O. 1999. Perceived air quality, sick building syndrome symptoms and productivity in an office with two different pollution loads. Indoor Air 9:165-179.

Wargocki, P., Wyon, D.P., Matysiak, B., and Irgens, S. 2005. The effects of classroom air temperature and outdoor air supply rate on the performance of school work by children. In Proceedings of Indoor Air 2005. Beijing, China: Tsinghua University Press.

Weale, R.A. 1963. The Ageing Eye. London: HK Lewis and Company.

Weale, R.A. 1992. The Senescence of Human Vision. Oxford: Oxford University Press.

Weisel, C.P., Zhang, J., Turpin, B.J., Morandi, M.T., Colome, S., Stock, T.H., Spektor, D.M., Korn, L., Winer, A., Alimokhtari, S., Kwon, J., Mohan, K., Harrington, R., Giovanetti, R., Cui, W., Afshar, M., Maberti, S., and Shendell, D. 2005. Relationship of Indoor, Outdoor and Personal Air (RIOPA) study: Study design, methods and quality assurance/control results. J. Expo. Anal. Environ. Epidemiol. 15(2):123-137.

Weiss, C.H. 1997. Evaluation: Methods for Studying Programs and Policies. Upper Saddle River, NJ: Prentice Hall.

Weschler, C.J., 2004. New directions: Ozone-initiated reaction products indoors may be more harmful than ozone itself. Atmos. Environ. 38:5715-5716.

Weschler, C.J., and Shields, H.C. 1997. Potential reactions among indoor pollutants. Atmos. Environ. 33:3487-3495.

Weschler, C.J., and Shields, H.C. 2000. The influence of ventilation on reactions among indoor pollutants: Modeling and experimental observations. Indoor Air 10:92-100.

Weschler, C.J., and Wells, R. 2004. Indoor chemistry and health effects. Indoor Air 14:373-375.

Weschler, C.J., Hodgson, A.T., and Wooley, J.D. 1992. Indoor chemistry: Ozone, volatile organic compounds and carpets. Environ. Sci. Technol. 30:3250-3258.

Weschler, C.J., Naik, D.V., and Shields, H.C. 1998. Indoor ozone exposures resulting from the infiltration of outdoor ozone. Pp. 83-100 in Indoor Air Pollution, Radon, Bioaerosols and VOCs. Chelsea, MI: Lewis.

Williams, G.M., Linker, H.M., Waldvogel, M.G., Leidy, R.B., and Schal, C. 2005. Comparison of conventional and integrated pest management programs in public schools. J. Econ. Entomol. 98(4):1275-1283.

Williams, N., and Carding, P. 2005. Occupational Voice Loss. New York: Taylor and Francis Press.

Winther, B., Gwaltney, J.M., Jr., Mygind, N., and Hendley, J.O. 1998. Viral induced rhinitis. Am. J. Rhinol. 9:17-20.

Wolkoff, P., Wilkins, C.K., Clausen, P.A., and Nielsen, G.D. 2006. Organic compounds in office environments—sensory irritation, odor, measurements and the role of reactive chemistry. Indoor Air 16:7-19.

WSBE (Washington State Board of Education). 2005. Washington High Performance School Buildings: Report to the Legislature. Seattle, WA: Paladino and Company.

Wu, C.C., and Lee, G.W.M. 2004. Oxidation of volatile organic compounds by negative air ions. Atmos. Environ. 38:6287-6295.

Wyon, D.P. 2004. The effects of indoor air quality on performance and productivity. Indoor Air 14(Suppl 7):92-101.

Yu, I.T., Li, Y., Wong, T.W., Tam, W., Chan, A.T., Lee, J.H., Leung, D.Y., and Ho, T. 2004. Evidence of airborne transmission of the severe acute respiratory syndrome virus. N. Engl. J. Med. **350**(17):1731-1739.

Zaragoza, M., Salles, M., Gomez, J., Bayas, J.M., and Trilla, A. 1999. Handwashing with soap or alcoholic solutions? A randomized clinical trial of its effectiveness. Am. J. Infect. Control 27:258-261.

Zimmer, R., and Toma, E. 2000. Peer effects in private and public schools across countries. J. Policy Anal. Manage. 19(1):75-92.

Appendix

Biographies of Committee Members

John D. Spengler, ***Chair,*** is the Akira Yamaguchi Professor of Environmental Health and Human Habitation in the Department of Environmental Health at Harvard University's School of Public Health. He has conducted research in the areas of personal monitoring, air pollution health effects, aerosol characterization, indoor air pollution, and air pollution meteorology. More recently, he has been involved in research that includes the integration of knowledge about indoor and outdoor air pollution as well as other risk factors into the design of housing, buildings, and communities. He uses the tools of life-cycle analysis, risk assessment, and activity-based costing as indicators to measure the sustainable attributes of alternative designs, practices, and community development. He serves as adviser to the World Health Organization on indoor air pollution, personal exposure, and air pollution epidemiology, and he has served as either a member or consultant on various U.S. EPA Science Advisory Board committees. He received a B.S. in physics from the University of Notre Dame, an M.S. in environmental health sciences from Harvard University, and a Ph.D. in atmospheric sciences from the State University of New York-Albany.

Vivian E. Loftness, ***Vice Chair,*** is a professor of architecture at Carnegie Mellon University. She is an international energy and building performance consultant for commercial and residential building design and has researched and written extensively on energy conservation, passive solar design, climate, and regionalism in architecture. Professor Loftness has worked for many years with the Architectural and Building Sciences

Division of Public Works Canada, researching and developing the issues of total building performance and the field of building diagnostics. Through the Advanced Building Systems Integration Consortium, she has been actively researching and designing high-performance office environments. Professor Loftness is a member of the advisory board at Johnson Controls and Owens-Corning Fiberglas Construction. She has been involved in international organizations including as resident, Executive Interchange Canada Architecture and Building Sciences, Department of Public Works Canada (October 1982 to June 1985) and as principal investigator, World Meteorological Organization, for Climate/Energy Graphics, and she has previously served on NRC committees. She holds a B.S. in architecture and a master's of architecture from the Massachusetts Institute of Technology.

Charlene W. Bayer is the Research Institute Branch head, principal research scientist, and adjunct professor at the Georgia Institute of Technology. A reviewer for the *Journal of Chromatographic Science, Indoor Air,* and *Analytical and Bioanalytical Chemistry,* and periodically for a variety of other peer-reviewed journals and conference proceedings, she has expertise in indoor environments, air quality, and related health concerns. Dr. Bayer is a past member of the ASHRAE Environmental Health Committee and serves on numerous other review committees for Underwriters' Laboratories, the U.S. Environmental Protection Agency, Oak Ridge National Laboratory, and the American Chemical Society. Dr. Bayer was a semifinalist in *Discovery Magazine* awards for Innovative Technology of Importance in 1999 and holds numerous patents for materials and devices for monitoring and improving air quality. Her current research includes a project to develop a personal monitoring vest able to monitor a variety of pollutants that are suspected as being asthmatic aggravators while linking these with pulmonary function tests sponsored by the Department of Housing and Urban Development. She holds a B.S. in chemistry from Baylor University and an M.S. and a Ph.D. in organic chemistry from Emory University.

John S. Bradley is a principal research officer at the National Research Council of Canada's Institute for Research in Construction. Dr. Bradley is involved in the design of efficient procedures for making advanced acoustical measurements in rooms for speech and music, and the use of these quantities to evaluate such spaces more scientifically; measuring techniques for predicting and evaluating speech intelligibility in rooms, including school classrooms; noise control related to buildings, including for outdoor noises; and relationships between physical and subjective assessments of annoyance caused by noise from various sources. He is a fellow in the Acoustical Society of America, past president of the Canadian

Acoustical Association, and a member of the Acoustical Society of America Technical Committee on Architectural Acoustics and the editorial board of the Audio Engineering Society. He has served on ANSI and ISO standards committees as well as the WHO working group for community noise guidelines. He holds a B.S. in physics and a master's in physics/acoustics from the University of Western Ontario, and a Ph.D. in physics/acoustics from Imperial College, University of London.

Glen I. Earthman is professor emeritus of educational administration at Virginia Polytechnic Institute and State University. His research interests extend to all phases of school facilities, with a concentration on exploring the relationship between school building condition and student achievement. Dr. Earthman has 40 years of experience in the field of education, serving as a teacher, principal, and executive director for school facility planning in the Philadelphia public schools, where he directed a staff of 250 professional planners and architects engaged in all activities associated with planning school facilities and monitoring the construction and evaluation of the resultant buildings. He is a member and past officer of the International Society for Educational Planning and the Council of Educational Facility Planners, International (CEFPI). He received the CEFPI President's Award for planning activities in 1992 and the Planner of the Year Award in 1994. He holds a B.A. and a master's degree from the University of Denver and a Ph.D. from the University of Northern Colorado, where he served as a graduate fellow in the School Planning Laboratory.

Peyton A. Eggleston is the director of the Center for Children's Environmental Health at Johns Hopkins University, a center of excellence sponsored by the National Institute of Environmental Health Sciences and one of EPA's National Centers for Environmental Research. He is also a professor of pediatrics at the Johns Hopkins School of Medicine. His research focus is environmental allergens—their role in respiratory diseases (in particular, asthma), risk factors for sensitization, means of avoidance, and methods and effectiveness of indoor environmental control. He is credited with more than 190 publications and serves on the editorial board of the Journal of *Allergy and Clinical Immunology*. He has served as a member of the Board of Allergy and Immunology and is an active member of the Academy of Asthma, Allergy, and Immunology. Dr. Eggleston received his medical degree from the University of Virginia and training in pediatrics and allergy-immunology at the University of Washington.

Paul Fisette is the director and an associate professor of building materials and wood technology and an associate professor of architecture at the University of Massachusetts, Amherst. Professor Fisette's research and professional focus involve the performance of building systems, energy-efficient construction, sustainable building practices, and the performance of building materials. He has developed an innovative Web service that provides technical advice on the performance, specification, and use of building materials. Professor Fisette has written more than 200 published works on building science and construction technology. Previous to his current position, he owned and operated a general contracting business and was a senior editor with *Custom Builder Magazine*, covering technical information and innovations of interest to small- and medium-sized residential building firms. Professor Fisette is a member of the National Research Council's Board on Infrastructure and the Constructed Environment (BICE) and a contributing editor for *The Journal of Light Construction*, and he has served on a variety of editorial and professional advisory boards. He holds B.S. and M.S. degrees in wood science and technology from the University of Massachusetts.

Caroline Breese Hall, M.D., is a professor of pediatrics and medicine in infectious diseases at the University of Rochester School of Medicine. At Rochester her research has focused on virology, especially respiratory syncytial virus, human herpes virus 6, and vaccines, resulting in more than 500 published articles. Among the national positions Dr. Hall has held are president of the Pediatric Infectious Diseases Society; member of the Red Book Committee for 8 years and chairman for 4 years; and member of the CDC's Advisory Committee of Immunization Practices, the Board of Scientific Counselors for the National Center of Infectious Diseases, committees for the Institute of Medicine, the American Board for Pediatric Infectious Diseases, and the Subboard for Pediatric Infectious Diseases. Among the awards she has received are the Pediatric Infectious Diseases Society Distinguished Physician Award and the Clinical Virology Award from the Pan American Society of Virology, as well as being named among the best doctors in America and among the top 20 women physicians in America. She graduated from Wellesley College and Rochester Medical School and did her subsequent residency training at Yale, followed by fellowships first in pediatric infectious diseases and then allergy and immunology in the Department of Medicine at Yale University.

Gary T. Henry is a professor of policy studies in the Andrew Young School of Policy Studies at Georgia State University, where he specializes in educational policy, school accountability, and program evaluation. He previously served as the director of evaluation and learning services for the David and Lucile Packard Foundation. Dr. Henry has evaluated a variety of policies and programs, including Pre-K and the HOPE Scholarship program in Georgia as well as school reforms and accountability systems. He has served as the director of the university's Applied Research Center. He is the author of *Practical Sampling* (Sage 1990) and *Graphing Data* (Sage 1995) and co-author of *Evaluation: An Integrated Framework for Understanding, Guiding, and Improving Policies and Programs* (Jossey-Bass 2000), and he has published extensively in the field of evaluation and policy analysis. In addition, he served as deputy secretary of education for the Commonwealth of Virginia and chief methodologist with the Joint Legislative Audit and Review Commission for the Virginia General Assembly. He received the Evaluation of the Year Award from the American Evaluation Association in 1998 for his work with Georgia's Council for School Performance and the Joseph Wholey Distinguished Scholarship Award in 2001 from the American Society for Public Administration and the Center for Accountability and Performance. He received his Ph.D. from the University of Wisconsin.

Clifford S. Mitchell is an associate professor and director of the Occupational Medicine Residency Program at the Johns Hopkins Bloomberg School of Public Health. His research interests include indoor air quality and its effects on human health in schools and in office buildings. He holds a B.A. from Williams College, an M.S. from the Massachusetts Institute of Technology, an M.D. from Case Western Reserve, and an M.P.H. from Johns Hopkins University, School of Public Health.

Mark S. Rea is the director of the Lighting Research Center at Rensselaer Polytechnic Institute (RPI), a position he has held since 1988. He is also a professor at the School of Architecture and in the Department of Philosophy, Psychology and Cognitive Science at RPI. Prior to RPI he was the manager of the Indoor Environment Program, Building Performance Section at the National Research Council of Canada. He also has been a visiting scientist at the Electricity Council Research Centre, Capenhurst, United Kingdom. He is a fellow of the Illuminating Engineering Society of North America and of the Society of Light and Lighting (United Kingdom). He is on the international editorial advisory board of the *Lighting Research and Technology Journal* and is editor-in-chief of the *Illuminating Engineering Society of North America Lighting Handbook* (8th and 9th editions). He received the William H. Wiley Distinguished Faculty Award from RPI and the Gold Medal from the Illuminating Engineering Society of North America. Dr. Rea received a B.S. in psychology and an M.S. and a Ph.D. in biophysics from Ohio State University.

Henry Sanoff is professor emeritus of architecture at the North Carolina State University College of Design. He came to the College of Design in 1966 from the University of California, Berkeley, where he was an assistant professor. He is a member of the Academy of Outstanding Teachers and has been designated an Alumni Distinguished Graduate Professor. Mr. Sanoff teaches courses related to community participation, social architecture, design research, design methodology, and design programming. He has been a visiting lecturer and scholar at more than 85 institutions in the United States and abroad. He is the U.S. editor of the *Journal of Design Studies* and a member of the editorial board of the *Journal of Architecture and Planning Research.* Professor Sanoff is also recognized as one of the founders of the Environmental Design Research Association (EDRA) in 1969. His research has concentrated on the areas of social housing, children's environments, community arts, aging populations, and community participation. Professor Sanoff received a bachelor of architecture and a master of architecture from Pratt Institute.

Carol H. Weiss is the Beatrice B. Whiting Professor of Education at the Harvard Graduate School of Education, where she teaches in the areas of administration, planning, and social policy. Her courses include evaluation methods, research methods, use of research as a strategy for change, and organizational decision making. She has published 11 books, 3 of which are on evaluation and 5 on the uses of research and evaluation in policy making. Her recent work is about the influences on educational policy making exerted by information, ideology, interests, information, and institutional constraints. She has been a fellow at the Center for Advanced Study in the Behavioral Sciences, a guest scholar at the Brookings Institution, a congressional fellow under the sponsorship of the American Sociological Association, a senior fellow at the U.S. Department of Education, and a member of seven panels of the National Academy of Sciences. She is on editorial boards for *Teachers College Record*, the *Journal of Educational Change, Journal of Comparative Policy Analysis, Asia-Pacific Journal of Teacher Education and Development, American Behavioral Scientist*, and others. She holds a Ph.D. in sociology from Columbia University.

Suzanne M. Wilson is a professor of teacher education in the Department of Teacher Education, Michigan State University. She is an educational psychologist with an interest in teacher learning and teacher knowledge. Her studies include the capacities and commitment of exemplary secondary school history and mathematics teachers, and she has written extensively on the knowledge base of teaching. She recently concluded a longitudinal study of the relationship between educational policy and teaching practice by examining efforts to reform mathematics teaching in California. She is also the director of the Center for the Scholarship of Teaching. Her areas of expertise include curriculum policy, history of teachers and teaching, mathematics reform, teacher assessment, teacher education and learning, and teaching history. Dr. Wilson earned her Ph.D. from Stanford University.